Fads and Facts about Vitamin D

Edited by Edward T. Zawada Jr.

Published in London, United Kingdom

IntechOpen

Supporting open minds since 2005

Fads and Facts about Vitamin D
http://dx.doi.org/10.5772/intechopen.79070
Edited by Edward T. Zawada Jr.

Contributors
Ishiaq Omotosho, Rodrigo Nolasco Santos, Marise Lazaretti-Castro, Binit Vaidya, Shweta Nakarmi, Julia Fedotova, Amir Khoshvaghti, Edward T. Zawada Jr.

First published in London, United Kingdom, 2019 by IntechOpen
IntechOpen is the global imprint of INTECHOPEN LIMITED, registered in England and Wales, registration number: 11086078, The Shard, 25th floor, 32 London Bridge Street
London, SE19SG - United Kingdom
Printed in Croatia

British Library Cataloguing-in-Publication Data
A catalogue record for this book is available from the British Library

Additional hard and PDF copies can be obtained from orders@intechopen.com

Fads and Facts about Vitamin D
Edited by Edward T. Zawada Jr.
p. cm.
Print ISBN 978-1-83881-143-3
Online ISBN 978-1-83881-144-0
eBook (PDF) ISBN 978-1-83881-145-7

For EU product safety concerns:
IN TECH d.o.o., Prolaz Marije Krucifikse Kozulić 3, 51000 Rijeka, Croatia, info@intechopen.com or visit our website at intechopen.com.

Meet the editor

Edward T. Zawada Jr., MD, was born in Chicago, Illinois, in 1947. He graduated with a BSc (summa cum laude) from Loyola University, Chicago, in 1969 and an MD (summa cum laude) from the Loyola Stritch School of Medicine in 1973. He was inducted into Alpha Omega Alpha in 1972. He trained in internal medicine and nephrology at the David Geffen School of Medicine at UCLA where he later served as an assistant professor of medicine. He had a long career in academic medicine at the University of Utah, Medical College of Virginia, and Sanford School of Medicine of the University of South Dakota where he is a professor and chairman emeritus of the Department of Internal Medicine. He is board certified by the American Board of Internal Medicine in internal medicine, nephrology, geriatrics, and critical care medicine. He also has board certifications in nutrition by the American College of Nutrition, pharmacology by the American Society of Clinical Pharmacology, and hypertension by the American Society of Hypertension. He has been awarded with a fellowship status by the American College of Physicians, the American College of Chest Physicians, the American Heart Association, the American Society of Nephrology, the American Geriatrics Society, the American College of Clinical Pharmacology, the Society for Vascular Medicine, and the Society of Critical Care Medicine. He was awarded a distinguished service award by the South Dakota Medical Society in 2002 for a career devoted to medical education and master of the American College of Physicians in 2005.

Contents

Preface

Vitamin D is the subject of great controversy. Initially, the controversy was based on the difficulty of obtaining accurate measurements of vitamin D. Then, subsequent discovery of multiple hydroxylated products led to confusion as to which are the most clinically important forms. Least controversial was the effect of vitamin D deficiency on bone formation. Finally, an explosion of reports on extraskeletal pathophysiology have suggested that vitamin D has a great impact on multiple tissues and organ function that can worsen diseases and affect mortality.

This book is organized into four sections: Section 1 "The Growing Importance of Vitamin D"; Section 2 "Extraskeletal Effects of Vitamin D Deficiency"; Section 3 "New Frontiers of Vitamin D Research"; and Section 4 "Clinical Application of Measurement of Vitamin D and Replacement." Chapters in these sections cover the earliest investigations of vitamin D, the extraskeletal consequences of vitamin D, and the impact of correction of vitamin D deficiency on overall health and mortality.

A unique theme developed in Section 2 is the similarity of vitamin D's structure to steroid molecules, including sex hormone structures. This is further discussed in the chapters on the effect of vitamin D on physical performance, its psychological impact on behavior, and its role in metabolism and response to stress. An important future area of research into both skeletal and extraskeletal consequences of vitamin D deficiency is a chapter on the study of vitamin D in space. Finally, clinical application of measurement and treatment is presented in the replacement of vitamin D to patients with rheumatologic problems and systemic problems such as those in the intensive care unit.

This book presents a concise summary of a broad body of international research that reports advantages as well as disadvantages of vitamin D in skeletal and extraskeletal tissue physiology.

Edward T. Zawada Jr. M.D., M.A.C.P.
Professor and Chairman Emeritus,
Department of Internal Medicine,
University of South Dakota,
Sanford School of Medicine,
Nephrologist and Intensivist,
Shasta Critical Care Specialists,
Redding, California

Section 1

Growing Importance of Vitamin D

Chapter 1

Introductory Chapter: Overview of Vitamin D Facts and Fads

Edward T. Zawada

1. Introduction

The importance of vitamin D has risen from a narrow focus on bone health to a major concern for individual and population survival due to possible extraskeletal pharmacologic effects. The latter is derived from alterations in vascular tone and multi-organ physiologic changes resulting from deficiency and restored by repletion. However, the literature is controversial because there are studies pro and con for these latter effects. Thus, there has arisen a need to try to dissect fact from fad.

I began my clinical and research interest in vitamin D measurement, physiology, and therapeutic replacement nearly 50 years ago while in training in nephrology at the University of California, Los Angeles (UCLA). It was as though the influence of Hollywood, the beach, and the star-laden music industry created a stereotype of young, beautiful, tanned, physically fit population that naturally lead to a concern with a vitamin which is activated by being outdoors, from the sunlight, from the exposure at beach. As it were, nephrology at UCLA at that time was still a newborn specialty descended from the original specialty called "metabolism." Gathered together at UCLA were many clinicians who were interested in vitamin D. Names such as Jack Coburn, Charles Kleeman, David B. N. Lee, Morton Maxwell, Nostratola Vaziri, Arnold Brickman, Francesco Llach, and Joel Koppel were those who lured me into the arena of calcium balance in health and disease. My first publication was a collaboration on a very detailed review of "The pathophysiology and clinical aspects of hypercalcemic disorders" [1]. Also developing at that same time was knowledge of the variety of hormonal and other factors that determined proper intake and serum levels of calcium, phosphorus, and vitamin D. Finally, much was being explored about the pharmacokinetics, pharmacodynamics, and excretion of these factors and the impact on bone and skeletal health.

Dialysis and transplantation would eventually emerge to consume the interest of subsequent nephrology trainees, but this was the time of the transition from the endocrinology of fluids and electrolytes controlled by the kidney and disturbed by kidney diseases to the explosion of kidney replacement strategies. The above predecessors attracted their like-minded academicians from around the world to come and lecture which included Shaul Massry, Eduardo Slatopolsky, Hector DeLuca, Ralph DeFronzo, and others. They often presented annually at the "vitamin D meeting" on the Monterey Peninsula. I was able to observe the dialog among these pioneers which filled my mind with knowledge but also questions for future investigation. What was most interesting about science that time was the indirect nature of proving that vitamin D and parathyroid hormone played such a variety of roles in kidney and other organ function and disease since both were not easily, if at all, measurable in those days. Bioassays and extractions from the liver and kidney and parathyroid glands were used in pharmacologic and dietary studies of electrolyte balance and organ physiologic measurements to search for perturbation of physiology, response to animal models of human disease, and observations intended to detect restoration of failed homeostasis.

2. Evolution of studies

By exposure to a large number of patients with calcium and vitamin D disorders, almost inadvertently I began to pursue a slightly different direction from my mentors. I was initially working in blood pressure management and prostanoid metabolism [2] when I observed a variety of hemodynamic disturbances in these "calcium" patients [3–9]. I questioned mechanisms and consequences of my observations. I began recording case reports and performing animal physiology studies and finally dietary studies in animals and then humans. The result was an idea that calcium and vitamin D influenced the heart, the blood vessels, and the physiology of circulation. This was the beginning of considerations of extraskeletal effects of calcium and vitamin D perturbations. Of course, in my mind I feel that I was in part responsible for discovering the impact of calcium and vitamin D on multi-organ, total body physiology, and homeostasis which confers survival benefit when deficiency is corrected and supplementation is provided. From my time in the 1970s, there has been a logarithmic accumulation of observations and research trials worldwide which has expanded the horizon of calcium and vitamin D metabolism from strictly bone health to multiple organ and total body health. I will review my original studies as well as the consequences of such hemodynamic and systemic effects of calcium and vitamin D in the last chapter of this book dealing with their role in the management of ICU patients.

3. Conclusion

Why is there a gap between the fads and facts with regard to vitamin D and health? For one thing, detailed pharmacokinetic measurements of vitamin D and its numerous metabolites in various populations are a relatively recent phenomenon. Secondly, it is likely that vitamin D plays a contributory role but not the primary role in disease and health beyond the known requirement to prevent rickets or osteomalacia, the main vitamin D-deficient bone diseases of children and adults, respectively. Finally, I believe more randomized controlled trials are needed to evaluate fad from fact.

Author details

Edward T. Zawada
University of South Dakota, United States

*Address all correspondence to: ezawada@sio.midco.net

References

[1] Lee DBN, Zawada ET, Kleeman CR. The pathophysiology and clinical aspects of hypercalcemic disorders. Western Journal of Medicine. 1978;**129**:278-320

[2] Kirschenbaum MA, Zawada ET. The role of prostaglandins in glucagon-induced natriuresis. Clinical Science. 1980;**28**:393-401

[3] Zawada ET, Brickman AS, Maxwell MH, Tuck M. Hypertension associated with hyperparathyroidism is not responsive to angiotensin blockade. Journal of Clinical Endocrinology and Metabolism. 1980;**50**:912-915

[4] Zawada ET, Bennett EB, Stinson JB, Ramirez G. The role of serum calcium in blood pressure regulation during dialysis. Archives of Internal Medicine. 1981;**141**:657-658

[5] Zawada ET, Bennett EP, Johnson M, Bennett D. The effects of changes in calcium on blood pressure in conscious dogs. Nephron. 1983;**33**:262-267

[6] Zawada ET Jr, Johnson M. Calcium chelation and calcium-channel blockade in anesthetized acute renovascular hypertensive dogs. Mineral and Electrolyte Metabolism. 1984;**10**:366-370

[7] Zawada ET Jr, TerWee J, McClung DE. Systemic and renal vascular responses to dietary calcium and vitamin D. Hypertension. 1986;**8**:975-982

[8] Gilliland M, Zawada ET Jr, McClung DE, TerWee J. Natriuretic effect of calcium supplementation in hypertensive women over forty. Journal of the American College of Nutrition. 1987;**6**:139-143

[9] Manson JE, Cook NR, Lee IM, et al. Vitamin D supplements and prevention of cancer and cardiovascular disease. New England Journal of Medicine. 2019;**380**:33-44

Chapter 2

Plant Sources of Vitamin D and Its Medicinal Application in Sub-Sahara Africa

Ishiaq Omotosho

Abstract

Vitamin D is one of the fat-soluble vitamins structurally derived from cholesterol and similar to the reproductive hormones by possessing the ABCD hexagonal rings. Its open seco-B triene B ring along with the possession of 1,25 dihydroxycholesterol side chain has been postulated as possible configuration allowing for its flexibility to interact with many tissues including bones causing myriads of biological effects that are medicinally beneficial. Aside from the established role of vitamin D in bone and mineral metabolic diseases, exploiting this property of the vitamin had been in practice in orthodox medical practice and even in traditional sub-Sahara method of managing many disease conditions.

Keywords: plant sources of vitamin D, medicinal application, bone and mineral metabolism

1. Introduction

1.1 Preamble

"Africa and by extension her sub-Sahara regions are noted for various folklores some of which center on application of various traditional methods and means in solving many problems including medical issues. It is thus a common saying in Africa that 'before the advent of tooth brush and paste, Africans have their own mouth cleaning devises".

2. Definitions

2.1 Vitamins

Vitamins are organic compounds essential in small quantities for normal physiologic and metabolic functioning of the body [1]. Vitamins are a large family usually grouped as micronutrients; they are mostly derived from diets with few exceptions and unlike other food substances, this family of micronutrients usually exist in complexes with one another and thus cannot be obtained from a single dietary source.

Functionally, vitamins are involved in various metabolic processes where they serve usually as coenzymes in various biochemical reactions associated with proper functioning of the whole organism [2]. They thus catalyze organic reactions by participating in the formation of hormones, cells, chemical structures of the nervous system, composition of genetic material and a host of other biological processes. They also combine with proteins to form enzymes which participate in various body reactions including in the development of body's immune system.

Probably because vitamins are present in small quantities, in the past, diseases of vitamin deficiencies were treated using various vitamins supplementarily in their management; however, advancement in science has led to many biochemical and biological methods that are appropriately used in the identification, measurement and diagnosis of diseases associated with many of the known vitamins.

Due to the involvement of these vitamins in several metabolic processes in spite of their small quantities, their deficiencies usually manifest clinically in various forms; for example, pellagra and beriberi are clinical conditions associated with the deficiency of niacin and thiamine respectively, (sub-groups of vitamin B), scurvy is a clinical condition associated with vitamin C deficiency, while osteomalacia (in adults) and rickets (in growing children) are associated with vitamin D deficiencies. Night blindness is associated with vitamin A deficiency, deficiency of vitamin B12 or that of folic acid is associated with megaloblastic anemia while spinal bifida has been associated with the deficiency of folic acid in the mother while the baby was in-utero [2]. Because vitamins as essential nutrients are mostly derived from diets, etiology of hypovitaminosis has always been associated with either in-adequate intake from diet or abnormality of absorption whereby a large quantity of these essential nutrients remains unabsorbed even when present abundantly in diet. Generally, vitamin toxicity is associated with the fat-soluble series because of their insolubility in the aqueous medium which largely constitutes the human system; however, excessive ingestion of vitamin A has been known to result in toxic manifestations which may ultimately result in liver damage [2].

2.2 Classification of vitamins

Vitamins are generally classified based on their solubility in aqueous or lipid medium. Thus, there are the fat-soluble vitamins (A, D, E and K) and the water-soluble ones (vitamins B complex and C). Although they are usually classified into these two broad groups, the classification is for convenience based on the chemical structure of vitamins; most of the vitamins have sub-groups that are no less prominent in name as the known main group. For example, while vitamin A is available either as the plant (carotenoid) or animal (retinol) types, vitamin B has several sub groups which have been distinctly classified based on structure and function. On the other hand, the name vitamin E, which is a known fat-soluble vitamin, refers to a family of eight naturally occurring homologs that are synthesized by plants from homogentisic acid. They are all derivatives of 6-chromanol and differ in the number and position of methyl groups on the ring structure. Also, vitamin D is another classical example of a lipid-soluble vitamin. The name, vitamin D, refers to about 5 different compounds generally classified as such, these are: 1,25-dihydroxyvitamin D3 (1,25-(OH)2D3, 24,25-dihydroxyvitamin D3 (24R,25-(OH)2D3, 1,25-dihydroxyvitamin D2 (1,25-(OH2)D), 25-hydroxyvitamin D3 (25-OH-D3), 25-hydroxyvitamin D2 (25-OH-D2). Hence, the general classification of vitamins notwithstanding, several sub-groups of vitamins exist which when considered based on structure, function and activity will make the conventional broad classification of vitamins too simplistic in terms of their overall relevance in human metabolic processes. The focus of this chapter is however on vitamin D with particular emphasis on its plant sources and medicinal applications in sub-Sahara Africa.

3. Vitamin D

Vitamin D is one of the fat-soluble vitamins that has distinct biochemical functions in human metabolism. It exists in five active forms

i. 1,25-dihydroxyvitamin D3(1,25-(OH)2D3,

ii. 24,25-dihydroxyvitamin D3(24R,25-(OH)2D3

iii. 1,25-dihydroxyvitamin D2(1,25-(OH2)D)

iv. 25-hydroxyvitamin D3(25-OH-D3)

v. 25-hydroxyvitamin D2(25-OH-D2)

3.1 Sources and production of vitamin D

The general conception is that vitamin D is synthesized only in the body, however, evidences abound that vitamin D is available in different forms in some plants and fruits in sub-Sahara Africa. There are also reports on application of these plant sources in the treatment of some vitamin D related diseases. Vitamin D2 (ergosterol) has been identified in some plants and fungi. Vitamin D2 differs from D3 in having a double bond between C22 and C23 and a methyl group at C24 in the side chain. D2 can be considered the first vitamin D analog which is converted to D3 by ultraviolet radiation. As earlier stated, plants like perennial ryegrass contain some amounts of ergosterol which when ingested can also be readily converted to D3 in the body [3].

Vitamin D3 has many dietary sources. The parent compound (D2) is derived essentially from dietary sources like egg yolk, sea fatty fish, liver, and mushroom among others. The production of vitamin D3 (D3) in the skin is not an enzymatic process. Sea fatty fish essentially contain vitamin D2.

3.2 Plant sources of vitamin D

Accidental discovery of activation of some vegetables and crops by exposure to mercury lamp led to the identification of vitamin D2 in some inert foods like cottonseed, wheat and lettuce [4]. Later, vitamin D2 was identified from solutions of ergosterol irradiated with UV light in-vitro [5]. Hence, contamination of plants with fungi which has a high concentration of ergosterol led to the discovery of "plants contaminated with fungi" as veritable source of vitamin D2. This initial concept on the presence of vitamin D in plants however changed with the discovery of a type of calcium intoxication in grazing animals similar to that caused by vitamin D toxicity that consumed certain plants [6]. This was believed to be due to vitamin D3 or a metabolite of vitamin D3 present in the plants that stimulate calcium absorption producing hypercalcemia and deposition of calcium in soft tissue including aorta, heart, kidneys, intestines, and uterus [6].

Hence, while plants like *Solanum glaucophyllum* Desf. (*S. glaucophyllum*), *Cestrum diurnum* L. (*C. diurnum*) and *Trisetum flavescens* Beauv. (*T. flavescens*) were found to cause calcium intoxication similar to that caused by vitamin D toxicity in grazing animals, rats and even chickens in South America [7–9]; similar effects of plants ameliorating calcium intoxication were found in studies on chicken in Africa and her sub-Sahara using *Moringa oleifera* leaves [10]. Aside from above, provitamin D3 and vitamin D2 (7-dehydro cholesterol) have been identified in leaves of plants like the genus *Solanaceae* and in *S. lycopersicum*, *S. glaucophyllum* and *C. annuum* amongst others [11–15].

Like most plant and herbal preparations, there are lots of knowledge gap in the biochemical and physiological mechanism behind application of these plant materials as sources of nutrients and most importantly in the management of diseases. However, isolation and characterization of vitamin-like substances in some of the plant may allow for the assumption of most of the claimed empirical roles of these plants and herbal preparations (WHO, 2002) [16].

Ocimum gratissimum (**Figure 1**) is one of the plants commonly used (empirically) to treat open wounds in rural setting of sub-Sahara Africa. The phytochemical analysis of this plant is well documented [17]. The assumed theory is that at higher temperatures, leaves of this plant facilitate collagen and fiber formation on the open wound thus enhancing clotting and angiogenesis [18]. However, the direct effect traceable to vitamin D content and activity remain to be elucidated.

3.3 Chemistry and biochemistry of vitamin D

Structurally, vitamin D is derived from cholesterol, and it is related to the classical steroid hormones with their traditional A B C D hexagonal rings. However, its uniqueness is derived by its possession of both 1,25-dihydroxycholesterol side chain and a seco-B triene structure. This lack of a complete B-ring as in steroids thus allows for the molecule to be depicted in a non-steroidal, extended formation in contrast to a classical steroid hormonal configuration which may or may not possess a truncated side chain but all have fully in-tact A B C D steroid rings (**Figure 2**). Biochemical evidences have shown that these structural differences allow for

Figure 1.
Ocimum gratissimum also known as clove basil, African basil, and in Hawaii as wild basil.

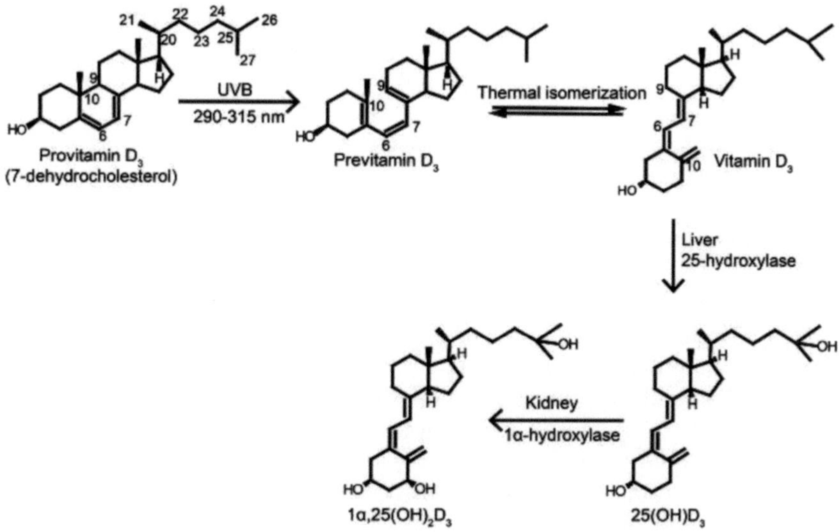

Figure 2.
Synthesis and activation of vitamin D. Vitamin D^3 is synthesized in the skin upon UVB exposure. The UVB exposure of provitamin D^3 (7-dehydro cholesterol) in the skin breaks the B-ring to form previtamin D^3, which undergoes thermally induced rearrangement to vitamin D^3. Vitamin D^3 is transported to the liver where it is hydroxylated at C-25 by the enzyme 25-hydroxylase producing 25OHD^3, which is the major circulating form in vertebrates. The 25OHD^3 is hydroxylated a second time at C-1 in the kidneys to the active metabolite 1,25 $(OH)^2D^3$. Figure adapted from Jäpelt et al. [15].

flexibility of these seco-B ring thus facilitating the interaction of 1,25 D3 with many tissues where specific biochemical interactions that produce biological effects occur [9].

In summary, vitamin D is made up of 4 rings: A, B, C and D, with a cyclopentanoperhydrophenanthrene nucleus. In contrast to the hormonal A B C D rings, the B ring has an open structure, there is a hydrocarbon chain attached to the rings at C17 position, a hydroxyl group (OH) at C3 while 4 methyl (CH3) groups are attached at C13, C18 and C25. In all, the molecule contains 27 carbon atoms. Variants of the molecule like ergocalciferol contain a double bond between C22 and C23 and additional methyl group at C24.

7-Dehydro-cholesterol (pro-vitamin D3) is photolyzed by ultraviolet light to pre-vitamin D3 which spontaneously isomerizes to vitamin D3 (**Figure 2**). Ultraviolet radiation with wavelengths between 290 and 315 nm causes the bond between the 9th and 10th position of the steroid ring to open, forming a compound called secosterol. This further undergoes cis-to-trans isomerization, by the formation of a trans-bond between the 5th and 6th carbon atoms, leading to the formation of vitamin D3, or cholecalciferol. Vitamin D3 (cholecalciferol) is converted into calcitriol (1,25 dihydroxycholecalciferol) by a process of hydroxylation in the liver and kidneys respectively, this is the active form of vitamin D. It is this very important involvement of ultraviolet radiation in the formation of active vitamin D that leads to its appellation as the “sunshine vitamin” [10].

3.4 Physiological effects of vitamin D

Physiologically, vitamin D can be broadly said to be essential for:

i. Regulation of calcium in the body

ii. Immune-modulation

As stated above, vitamin D is essentially associated with calcium regulation and immune modulation. Because of its effect on calcium absorption regulating mineral metabolism and bone growth, it is often referred to as a hormone. Physiologically, vitamin D facilitates intestinal absorption of calcium and also stimulates absorption of phosphate and magnesium ions. In the absence of vitamin D, efficiency of dietary calcium absorption is low. Furthermore, absorption of calcium is enhanced by vitamin D by stimulating the expression of a number of proteins involved in transporting calcium from the lumen of the intestine across the epithelial cells into the blood.

The indispensable role of vitamin D in calcium homeostasis and by extension bone metabolism has been demonstrated in many in-vitro animals' experiments. As a transcriptional regulator of bone matrix proteins, vitamin D induces the expression of osteocalcin and suppresses synthesis of type I collagen [11]. Also, in cell cultures, vitamin D stimulates differentiation of osteoclasts [12]. Although these changes may not be very crucial in humans, it's however been shown that the crucial effect of vitamin D on bone is to provide the proper balance of calcium and phosphorus to support mineralization [13]. An interesting finding was that vitamin D receptors are present in most if not all cells in the body. Additionally, experiments using cultured cells have demonstrated that vitamin D has potent effects on the growth and differentiation of many types of cells. These findings suggest that vitamin D has physiologic effects much broader than a role in mineral homeostasis and bone function. As one example, many immune cells not only express vitamin D receptors, but are capable of synthesizing active vitamin D, and deficiency in vitamin D has been associated with increased incidence of autoimmune disease and susceptibility to diseases [14]. This may be the basis of the uniqueness of the chemical structure of vitamin D molecule which, because of its seco-B triene on the open B ring, allows for its flexibility that facilitates interaction with many tissues to produce various biological effects [9]. It may therefore be inferred that it is this flexibility that probably allows though empirically for its association (in its crude form) and application in the management of many diseases in Africans natives. A brief description of some of these applications is listed below:

3.5 Diseases and medicinal applications of vitamin D

Diseases of vitamin D are usually associated with its deficiency which is often due to inappropriate dietary intake or in some climates and cultural settings, due to inappropriate exposure to ultra-violet light necessary to convert the ergosterol to the pre-vitamin format at the dermal level. Conventionally, circulating level of 25 hydroxy vitamin D is used in assessing appropriateness or other wise of vitamin D in the body. Based on this, vitamin D deficiency is defined by most experts as a level of less than 20 ng per milliliter (50 nmol per liter) [16–19, 24]. Hence, a level of hydroxy vitamin D $\geq$ 30 ng per milliliter is considered normal or adequate to affect the various functions of this vitamin [16].

Since vitamin D is normally synthesized in the body, diseases due to its deficiency have not been fully ascertained. However, certain health conditions in the elderly and in some climatic and cultural situations not allowing for adequate exposure to UV light have been described. Such disease conditions include:

3.5.1 Osteoporosis

This is a condition where bone density is decreased causing weakness in the bone and its inability to carry the frame of the body. This has been reported to be prevalent amongst women and approximately 33% of women aged between 60 and

70 and 66% of those over 80 have osteoporosis [20, 25]. Reduced bone mineral density increases the risk of fractures, which significantly contributes to morbidity and mortality of older persons [21, 26]. Physiologically, vitamin D deficiency is associated with reduced intestinal calcium absorption leading to an imbalance in calcium level and attendant reduction in bone mineral content resulting in reduced bone mineral density. Although there are few exceptions, generally, treatment of osteoporosis with vitamin D supplements have been reported severally underscoring and establishing the importance of vitamin D in bone mineral contents and the diseases associated with its imbalance [22, 27]. As stated above, essentially, osteoporosis is more prevalent among women especially at the post-menopausal stage. It's been shown that the occurrence may not be unconnected with a disruption of their hormonal profile which is occasioned by the menopause. However, natives in sub-Sahara Africa (and in other areas of the world like China) had used plants and herbs for the management of osteoporosis in this group of people. For example, the fruit *Avocado* (*Persea americana*) has been used for ages in mitigating many bone diseases including osteoporosis [23, 28]. Hence, several unpublished reports on the consumption of this fruit by native Chinese and Africans to alleviate the pains of menopausal changes abound.

Physiologically, oxidative and inflammatory stress have been adduced as the pathological basis of changes in bone that eventually result in osteoarthritis and osteoporosis, avocados fruit has been found to play important role in decreasing this oxidative and inflammatory stress due to the presence of large number of xanthophyll carotenoids such as lutein and zeaxanthin which are rich in anti-oxidants. Hence, consumption of this fruit by the natives in parts of the world where they are abundant has helped in mitigating the osteoporotic effect of hypocalcaemia occasioned by a deficiency in vitamin D. Other unpublished works have also reported the presence of other vitamins (vitamins B, C, K and E) and a high concentration of potassium and magnesium in the fruit.

3.5.2 Muscle weakness

Muscle weakness is also a prominent feature of vitamin D deficiency. Impaired muscle function with non-specific muscle weakness especially in the elderly has been documented in patients to cause falls. Discovery of vitamin D receptor in skeletal muscle tissue suggestive of possible vitamin D activity even in this region may be an indication of this vitamin's involvement in muscular activity and by inference it's possible ameliorative role in non-specific muscle weakness. Although, studies on this aspect of vitamin D in terms of dosage have been equivocal, the fact that supplementation with vitamin D in this condition has produced relief in some documented cases [24, 29] lend credence to the medicinal use of vitamin D in ameliorating non-numerical specific muscle weakness.

3.5.3 Hypertension

Oral supplementation with vitamin D is becoming a common practice in the treatment of hypertension in Blacks; this was because reduced level of 25 hydroxy vitamin D was found to be common in Blacks with diastolic hypertension. Although, the mechanism of this amelioration is yet to be fully understood, the practice is gradually gaining ground amongst Blacks [25, 30]. This may be the basis of consumption of several herbal preparations which mostly contain many algae, fruits and plants like *Ocimum gratissimum* in the empirical treatment of hypertension in African natives. Unfortunately, there are no published data or scientific investigations to support some of these activities.

3.5.4 Fixing of fractures

In sub-Sahara, it is not uncommon to see fractures being fixed by native doctors using the local technology. In most of such cases, the fractured bones are pieced together in their native format and incubated at fairly warm temperatures (40–50°) after wrapping with plants leaves such as wild *Ocimum gratissimum* [26, 31] such fractures often healed up after few days (7–21 days). Further studies on this traditional method of fixing fractures using broken/fractured chicken legs have confirmed the possible efficacy of this method in the local treatment of fractures [27, 32]. Although, most of these methodologies are empirical, the possibility of enhanced osteoblastic activity due to increased vitamin D supplementation from the leaves cannot be totally ruled out. We have also applied *O. gratissimum* experimentally in rats to ameliorate the debilitating effect of lead (Pb) toxicity probably based on the antioxidant and anti-inflammatory properties of this plant [28, 33].

3.5.5 Autoimmune disorders and diseases

Recently, some autoimmune disorders like rheumatoid arthritis and multiple sclerosis have been linked with vitamin D deficiency; an increasing number of this condition is responding to vitamin D supplementation [29, 34]. Although, the mechanistic effect is not yet known, possibility of proliferation of anti-inflammatory cytokine-Tumor Growth Factor (TGF)-induced by the upregulation of calcium level due to vitamin D supplementation in these subjects remain a probable hypothesis in the physiological role of vitamin therapy in this condition.

4. Conclusion

The multipurpose application of vitamin in the management of many diseases is not in doubt; however, the biochemical and physiological basis of its application in most of the disease states where it has been used either prophylactically or curatively are still under investigation. In the sub-Sahara, as stated in the prolog to this chapter, though empirical, medicinal use of leaves especially in fixing fractures has been and is still a popular practice especially in rural communities where access to basic medical facility is limited.

Author details

Ishiaq Omotosho
Department of Chemical Pathology, College of Medicine, University of Ibadan, Nigeria

*Address all correspondence to: iomotosho2014@gmail.com

References

[1] Anderson KN, Anderson LE. Mosby's Pocket Dictionary of Medicine, Nursing and Allied Health. 2nd ed. St. Louis, USA: Mosby Publishers; 1994

[2] Bishop ML, Fody EP, Schoeff LE. Clinical Chemistry. 7th ed. Philadelpia, USA: Lippincott Williams and Wilkins; 2013. pp. 686-688

[3] Japelt RB, Jakobsen J. Vitamin D in plants: A review of occurrence, analysis and biosynthesis. Frontiers in Plant Science. 2013;**4**:136

[4] Henry KM, Kon SK, Thompson SY, McCallum JW, Stewart J. The vitamin D activity of pastures and hays. The British Journal of Nutrition. 1958;**12**: 462-469

[5] Wallis GC, Kennedy GH, Fishman RH. The vitamin D content of roughages. Journal of Animal Science. 1958;**17**:410-415

[6] Mello JRB. Calcinosis—Calcinogenic plants. Toxicon. 2003;**41**:1-12

[7] Dallorso ME, Gil S, Pawlak E, Lema F, Marquez A. 1, 25(OH)2 vitamin D concentration in the plasma of Solanum glaucophyllum intoxicated rabbits. Australian Veterinary Journal. 2008;**79**: 419-423

[8] Weissenberg M, Maoz A, Levy A, Wasserman RH. Radioimmunoassay for rapid estimation of vitamin D derivatives in calcinogenic plants. Planta Medica. 1988;**54**:63-65

[9] Basudde CDK, Humphreys DJ. The vitamin D_3 metabolite–type activity of *Solanum malacoxylon*. Clinical Endocrinology. 1976;**5**:109-119

[10] Gopalakrishnan L, Doriya K, Kumar DS. *Moringa oleifera*—A review on nutritive importance and its medicinal application. Food Science and Human Wellness. 2016;**5**(2):49-56

[11] Curino A, Milanesi L, Benassati S, Skliar M, Boland R. Effect of culture conditions on the synthesis of vitamin D_3 metabolites in *Solanum glaucophyllum* grown *in vitro*. Phytochemistry. 2001;**58**:81-89

[12] Skliar M, Curino A, Milanesi L, Benassati S, Boland R. *Nicotiana glauca:* Another plant species containing vitamin D_3 metabolites. Plant Science. 2000;**156**:193-199

[13] Björn LO, Wang T. Is provitamin D a UV-B receptor in plants? Plant Ecology. 2001;**154**:1-8

[14] Jäpelt RB, Silvestro D, Smedsgaard J, Jensen PE, Jakobsen J. LC-MS/MS with atmospheric pressure chemical ionization to study the effect of UV treatment on the formation of vitamin D_3 and sterols in plants. Food Chemistry. 2011b;**129**:217-225

[15] Jäpelt RB, Silvestro D, Smedsgaard J, Jensen PE, Jakobsen J. Quantification of vitamin D_3 and its hydroxylated metabolites in waxy leaf nightshade (*Solanum glaucophyllum* Desf.), tomato (*Solanum lycopersicum* L.) and bell pepper (*Capsicum annuum* L.). Food Chemistry. 2012;**138**:1206-1211

[16] Michael F. Vitamin D deficiency. The New England Journal of Medicine. 2007;**357**:266-281

[17] Holick MF, Siris ES, Binkley N, Beard MK, Khan A, Katzer JT, et al. Prevalence of vitamin D inadequacy among postmenopausal North American women receiving osteoporosis therapy. The Journal of Clinical Endocrinology and Metabolism. 2005;**90**:3215-3224

[18] Lips P, Hosking D, Lippuner K, Norquist JM, Wehren L, Maalouf G,

et al. The prevalence of vitamin D inadequacy amongst women with osteoporosis: An international epidemiological investigation. Journal of Internal Medicine. 2006;**260**:245-254

[19] Thomas KK, Lloyd-Jones DM, Thadhani RI, Shaw AC, Deraska DJ, Kitch BT, et al. Hypovitaminosis D in medical inpatients. The New England Journal of Medicine. 1998;**338**:777-783

[20] Bischof-Ferrari HA, Wilet WC, Wong JB, Giovannucci E, Dietrich T, Dawson Hughes B. Fracture prevention with vitamin D supplementation: A meta analysis of randomized controlled trials. JAMA. 2005;**293**:2257-2264

[21] Lips P, Graafmans WC, Ooms ME, Bezemer PD, Bouter LM. Vitamin D supplementation and fracture incidence in elderly persons: A randomized, placebo controlled clinical trial. Annals of Internal Medicine. 1996; **124**:400-406

[22] Graafmans WC, Ooms ME, Hofstee HM, Bezemer PD, Bouter LM, Lips P. Falls in the elderly: A prospective study of risk factors and risk profiles. American Journal of Epidemiology. 1996;**143**:1129-1136

[23] Kune J, Tremblay J, Belavance F, Hamet P. Influence of environmental temperature on the blood pressure of hypertensive patients in Montreal. American Journal of Hypertension. 1991;**4**:422-426

[24] WHO IARC Working Group on the Evaluation of Carcinogenic Risks to Humans. Some Traditional Herbal Medicines, Some Mycotoxins, Naphthalene and Styrene. IARC Monogr Eval Carcinog Risks Hum. 2002;**82**:1-556

[25] Dike MC. Proximate phytochemical and nutrient compositions of some fruits, seeds and leaves of some plant species at Umudike, Nigeria. ARPN Journal of Agricultural and Biological Science. 2010;**5**(1):7-16

[26] Broe KE, Chen TC, Weinberg J, Bischof-Ferrari HA, Holick MF, Kiel DP. A higher dose of vitamin D reduces the risk of falls in nursing home residents: A randomized, multiple-dose study. Journal of the American Geriatrics Society. 2007;**55**:234-239

[27] Skodt H, Gallangher JA, Beresford JN, Couch M, Poser JW, Russell RG. Vitamin D metabolites regulate osteocalcin synthesis and proliferation of human bone cells in-vitro. The Journal of Endocrinology. 1985;**105**(3): 391-396

[28] Vivian Goldschmidt MA. Avoocado fruit in the treatment of bone diseases. 2010 (unpublished)

[29] Francesch RT, Romano PR, Park K-Y. Regulation of typ. 1 collagen synthesis by 1,25-dihydroxyvitamin D_3 in human osteosarcoma cells. Journal of Biological Chemistry. 1988;**263**(35): 18938-18945

[30] Krause R, Buhring M, Hopfenmuler W, Holick MF, Sharma AM. Ultraviolet B and blood pressure. Lancet. 1998;**352**: 709-710

[31] Owolabi AO, Alebiosu TK. Empirical use of *Ocimum gratisimum* traditionally in fixing fractures in a local setting. Trado-medical Practice. 2003;**4**: 114-120

[32] Gozdzialska A, Jaskiewicz J, Knapik-Czajka M, Drag J, Gawlik M, Ciesla M, et al. Association of calcium and phosphate balance, vitamin D, PTH, and calcitoninin patients with adolescent idiopathic scoliosis. Spine (Phila Pa 1976). 2016;**41**(8):693-697

[33] Omotosho IO, Henshaw AI, Adeagbo GA. Effect of natural antioxidant (*Ocimum gratissimum)* in modulating neurodegenerative changes

in rats fed with high concentration of lead acetate. Journal of Medicinal Plant Research. 2011;**5**(13):2743-2747

[34] Kriegel MA, Manson JAE, Costenbader KH. Does vitamin D affect risk of developing autoimmune disease?: A systemic review. Seminars in Arthritis and Rheumatism. 2011;**40**(6):512-531

Section 2

Extraskeletal Effects of Vitamin D Deficiency

Chapter 3

Vitamin D and Physical Performance: What Is the Ergogenic Actions of Vitamin D?

Rodrigo Nolasco and Marise Lazaretti-Castro

Abstract

Vitamin D produced in the skin by the action of the sun's rays turns into calcitriol, a powerful hormone, recognized as important for health. Although its most known effects are on mineral homeostasis and bone metabolism, its receptors (VDRs) have been identified in almost all tissues, suggesting that it should have other actions. Vitamin D acts directly on the skeletal muscle system maintaining muscle mass, strength levels and speed of muscle contraction. Thereby, allied to that, vitamin D is among the potential factors that are related to maintaining bone, and cannot be dissociated from the prevention of osteoporosis and sarcopenia. However, in the physical performance aspect, there are still uncertainness in the literature about the use of vitamin D as an ergogenic resource aimed at improving the physical performance of amateur and professional athletes. Therefore, due to the biological actions of vitamin D and high prevalence of low levels in sedentary and physically active individuals, this chapter will discuss the facts pointed out in the literature about the action of vitamin D as an ergogenic resource aiming at the preservation or improvement of the physical, including strength muscular, aerobic capacity and balance.

Keywords: vitamin D, athletic performance, lung function, sarcopenia

1. Introduction

The increase in human life expectancy during the previous century has raised new health issues, especially the control of aging-related deterioration. Important efforts have been made to gain new insights that may lead to modalities to delay the functional impairment and progression of chronic degenerative diseases, as well as sarcopenia and osteoporosis. Undoubtedly, the physical exercises are directly interrelated with the improvement or even rehabilitation of the physical performance besides increasing the life expectancy [1]. However, the effect of hormonal action, especially vitamin D, has lately been part of this prospect.

In this context, vitamin D has attracted considerable interest among health researchers, professional organizations and the lay public in recent times. Although it is called vitamin, conceptually vitamin D is a hormone. This is due to its hormonal nature, such as the ability to be integrally produced by the organism and to have specific receptors in several tissues [2].

Vitamin D has emerged for more than 500 million years. Even though its function in plants and invertebrates is unknown, the close association between vitamin D and sunlight has become essential in the evolution of terrestrial vertebrates. The main physiological function of vitamin D is to maintain the supply of calcium and phosphorus for complete mineralization of bone tissue [2].

Sunlight is the main source for producing the right amount of vitamin D for most humans. In food, it is found in small amounts, and there are few food sources. When it comes from sunlight, it is estimated that around 80–100% of the human needs for vitamin D come from exposure to sunlight [2, 3].

Although it is recognized as important for health, it is estimated that there are approximately 1 billion people in the world with inadequate concentrations of vitamin D. Furthermore, individuals with insufficient vitamin D levels have an increased risk of bone disease, such as: rickets, low bone mass and fractures due to increased bone resorption as a consequence of an overproduction of parathyroid hormone (PTH) [4].

Regarding functional capacity, the inadequacy of vitamin D stocks has catabolic effects on the musculoskeletal system, causing muscle weakness, lags in balance and impairs the formation of cross-bridges, which could impair physical performance. However, there remains divergences. Reports from recognized institutes do not corroborate the adequate levels for non-skeletal outcomes, or even the existence of evidence of a non-linear association for some results on physical performance [5].

Finally, recent studies have suggested a possible action of vitamin D in the lung function of individuals without lung disease. Similarly, results were observed on aerobic capacity being influenced by a possible action of vitamin D [6].

2. Vitamin D actions

Vitamin D is a steroid hormone. Its precursor found in animal tissues is 7-dehydrocholesterol, which is synthesized in the skin and is also the immediate precursor of cholesterol. Cholecalciferol has as its main source cutaneous synthesis catalyzed by ultraviolet light B (UVB). Its synthesis is initiated in the skin, under the action of UVB rays, which transform its precursor, 7-dehydrocholesterol, into cholecalciferol or vitamin D3. In smaller amounts, vitamin D can also be obtained through the diet of fortified foods such as dairy products and cereals, fatty fish and cod liver oil. After food intake or synthesis in the skin, vitamin D is transported into the liver where it is converted into 25-hydroxyvitamin D [25(OH) D] or calcidiol, which is the main form of circulating vitamin D and also used for serum dosage. In the kidney, 25(OH) D is converted to its biologically active form, 1,25 dihydroxyvitamin D (1,25 (OH) 2 D) [7, 8]. The biological actions of 1,25 (OH) 2 D are mediated by the nuclear transcription factor, called the vitamin D receptor (VDR) located in the cell nucleus [9].

VDR is part of the nuclear receptors of transcription factors regulating steroid hormones, retinoic acid, thyroid hormone, and vitamin D. Following the connection with VDR, there is the formation of the heterodimeric complex (Vitamin D Receptor—Retinoic Acid Receptor). This, in turn, binds to specific DNA sequences, also called the vitamin D responsive element (VDRE), promoting conformational changes that lead to the recruitment of several other transcriptional coactivators, resulting in the transcription of target genes [10].

After conversion to its biologically active form, 1,25-dihydroxyvitamin D regulates the expression of more than 900 gene variants [11]. Considering this aspect, it has been observed that these gene expressions have a significant impact on a range of variables related to health and performance, such as exercise-induced inflammation, pulmonary function, tumor suppressor genes, neurological function, cardiovascular health, glucose metabolism, health bone and skeletal muscle performance [12].

Vitamin D plays a key role in maintaining basal serum calcium and phosphorus levels for a variety of metabolic functions, regulation of transcription and bone metabolism. 1,25 (OH) 2D interacts with VDR in the small intestine to increase intestinal calcium absorption from 10–40% and phosphorus from approximately 60–80% [13]. Its action also extends to the cells responsible for bone remodeling, from the connection to the pre-osteoblasts, acting as a stimulus for precursor cells in osteoclasts from the RANK/RANK ligand system. Active osteoclasts remove calcium and phosphorus from bone to maintain serum levels of these elements. In the kidneys, 1,25 (OH) 2D stimulates calcium reabsorption of glomerular filtrate [3, 13]. Vitamin D and calcium are among the potential factors related to maintaining bone and muscle health, and cannot be separated from osteoporosis prevention in postmenopausal women [14].

2.1 Vitamin D and muscle

Moreover to the effects on bone metabolism, studies over 80 suggest improvements in physical performance in individuals exposed to UV radiation. Although these studies do not directly describe the action of vitamin D, induced changes in vitamin D levels may have played a role in muscle function [15].

Especially in older women with low vitamin D status, several intervention studies have reported that vitamin D supplementation increases appendicular muscle strength and improves physical function. In the musculoskeletal system, vitamin D exerts specific receptor-mediated functions (VDR) in processes ranging from protein synthesis to kinetics of muscle contraction, directly affecting the functional capacity of postmenopausal women [16]. Apart from this, the important mechanisms by which vitamin D can exert on human skeletal muscle can be classified as genomic or non-genomic [15].

Considering the genomic theory, it describes that 1,25 (OH) 2 D exerts a direct effect on the human muscular VDR, sparking progressive epigenetic changes that may have an impact on the morpho functionality of skeletal muscle. Within this context, in a study of women with limited mobility and with a relatively low level of vitamin D, vitamin D3 supplementation resulted in a 30% increase in intramuscular VDR protein concentration and a 10% increase in total muscle fiber size I and II. These findings corroborate the hypothesis that vitamin D contributes to the muscle mass of individuals with a tendency towards functional disability [17].

Muscle tissues have specific nuclear receptors for 1,25-[OH] 2D. In patients with strokes with atrophied type II fibers, improvement after vitamin D supplementation was observed for 25-OHD serum deficient patients before therapy. Also, improvements in muscle strength on the intact side of these patients with vitamin D-supplemented strokes were observed. In addition, in cross-sectional analysis, there was a correlation between 25-OHD and fiber diameter and type II [18].

Conversely, the non-genomic hypothesis credits a rapid and indirect mechanism by which 1,25 (OH) 2 D activates a series of secondary messenger processes that promote increased calcium kinetics. Researchers have extensively searched the so-called 1,25 (OH) 2 D non-genomic activities through the investigation of membrane proteins and intracellular signaling [19]. They have demonstrated that 1,25 (OH) 2 D can be mediated by a different membrane-associated rapid response steroid binding protein (MARRS) to facilitate rapid responses. It has been found that this protein is similar to the multifunctional isomerase disulfide protein of family A, member 3 (PDIA3), an endoplasmic reticulum enzyme. Interestingly, the antibody that blocks this protein prevented the transport of calcium and phosphate through the membranes of the intestinal epithelial cells. Therefore, skeletal muscle functionality can be influenced by intracellular effects on calcium handling through the action of vitamin D [20].

In addition, the presence of the VDR in vascular tissue and cardiac muscle seems to support the hypothesis that the vitamin D may impact the cardiovascular system's ability to transport oxygenated blood and the ability of skeletal muscles to use oxygen [21]. In a randomized double-blind placebo-controlled study with postmenopausal women, was observed a 23.5% increase in muscle strength and a reduction in the number of falls by 76% after 9 months of vitamin D supplementation [22]. In another study with older adults, significant improvements in physical performance, specifically in the up-and-go test, were reported with 2000 IU of vitamin D per day in 300 elderly women with a baseline level of 25 OHD below 24 ng/mL (60 nmol/L), [23].

In relation to the adequacy of vitamin D levels for individuals with low bone mass, this population seems to benefit in the physical performance, by serum concentrations of 25 (OH) D from 30 ng/mL (75 nmol/L) concentrations close to 36 ng/mL (90 nmol/L) as the most advantageous. Likewise, these 25 (OH) D values seem to benefit lower limb muscle strength, which was assessed by the walking test. Individuals with 25 (OH) D concentrations between 36 and 40 ng/mL (90 and 100 nmol/L) appear to perform the test faster [24].

Considering the adequacy of vitamin D levels, researchers in prospective, double-blind, placebo-controlled, randomized trial included Brazilian people institutionalized that received a 6-month supplementation of vitamin D, had as result, the increase in 16.4% in their maximum isometric strength of hip flexors and 24.6% in knee extensors measured by a portable mechanical dynamometer at 6 months, nevertheless the calcium/placebo group showed no improvement at all [25]. The same way, in other study, researchers have found an increase in neuromuscular parameters such as the balance (4.5%), functional mobility (10.1%) and muscle strength (5.7%) after elderly supplementation with vitamin D (6 month), without the regular practice of physical activity, considering that after the study most subjects reached the sufficiency level [26].

Vitamin D supplementation in youngsters has also been shown to be effective on the muscular strength of dancers who received oral supplementation of 2000 IU/day of vitamin D3 for 4 months during the winter. At the end of the study, the supplemented group presented increased isometric strength (18.7%, $p < 0.01$), plyometry (7.1%, $p < 0.01$), and reduction of lesions when compared to control ($p < 0.01$) [27]. On the other hand, in a study with 179 vitamin D-deficient Lebanese adolescents, vitamin D supplementation did not show improvement in manual grip strength [28].

Even with some intriguing results, studies are still conflicting about the action of vitamin D on muscle performance. Some meta-analyzes have found antagonistic outcomes. The meta-analysis who analyzed 17 randomized controlled trials in individuals of all ages including younger subjects just showed benefit in muscle strength in subjects with vitamin D serum levels below 25 nmol/L at baseline. In another hand, the meta-analysis which reviewed results from 13 randomized controlled trials in individuals older than 60 years old, observed a small benefit of daily vitamin D supplementation (800 IU–1000 IU per day) for balance and muscle strength [29].

2.2 Vitamin D actions on the respiratory system

The interrelationships between vitamin D metabolism and respiratory function have been studied in the literature, mainly in diseases of the respiratory tract, however the results are still not conclusive. Accordingly, the mechanisms by which life D could affect lung function have not yet been fully elucidated. However, some explanations have been pointed out. It is believed that vitamin D influences approximately 3% of the human genome and directly or indirectly, vitamin D controls many genes that are involved in the regulation of cell proliferation, differentiation and apoptosis of healthy and pathological cells [30].

1,25 (OH) 2D is mostly derived from the kidneys, however, other tissues, including the breast and prostate, may hydroxylate vitamin D in its active form. Activated, this metabolite is transported throughout the circulation by additional vitamin D binding proteins and lipoproteins; a function that allows vitamin D to actuate a wide range of skeletal and extra-skeletal functions [31].

Studies on the action of vitamin D on different cell types suggest that 1,25 (OH) 2D modulates smooth muscle excitation-contraction via intracellular Ca2+ release and Ca2+ sensitization. Airway resistance is dictated largely by the diameter of these pathways, and minor changes in this structure can significantly increase airway resistance. The absence of Vitamin D could affect the diameter of the airways, impairing pulmonary function [31, 32].

Most nucleated cells express the VDR, however the expression varies according to cell specificity. In the lungs, VDR was found in smooth muscle cells of the airway and in alveolar cells, also known as pneumocytes [32–34]. The enzyme 1α-hydroxylase, responsible for the conversion of 25 (OH) D into its active form, is also expressed by tracheal and bronchial cells [35].

Nguyen and cols, demonstrated through a series of studies with rat fetuses that type II alveolar cells underwent 1,25 (OH) 2D 3 action and suggested that vitamin D is important for maturation and the production of surfactants [36, 37]. Although in humans the mechanisms are more complex than in rats, the effect of vitamin D on the production of surfactants has been confirmed [33].

In relation to the growth and pulmonary maturation, studies with rats born to mothers with dietary vitamin D showed a loss of lung compliance compared to those born to mothers whose vitamin D supplementation [38]. In a similar study, vitamin D-deficient mice decreased lung volumes compared to the offspring of vitamin D-rich mice [39] .

In human studies, the authors used the presence of calbindin, a vitamin D-dependent calcium binding protein, as a molecular marker of the action of 1,25 (OH) 2 D3 on tissues. The authors found high levels of calbindin in fetal lung tissue between the 14th and 32nd weeks of gestation, suggesting an action of vitamin D on lung development [40].

In a cross-sectional study from the Third National Health, Nutrition and Examination Survey (NHANESIII) with a sample of 14,000 Americans, researchers found a positive correlation between FEV 1 and FVC with serum 25 (OH) D levels. The authors also observed that adults with low serum 25 (OH) D levels had lower than predicted FEV 1 of pulmonary function [41].

Although vitamin D intervention studies in postmenopausal women aiming for respiratory capacity are still scarce, in our study center, we developed a research project to evaluate the effects of a differential aquatic exercise program (HYDROS) on the musculoskeletal system in postmenopausal women compared to sedentary controls [26]. The large volume of parameters obtained during the six-month study was presented as an opportunity to evaluate, in a post-hoc analysis, the effects of vitamin D supplementation, combined or not with high intensity aquatic exercises, on the pulmonary function of postmenopausal women. We observed an improvement in the spirometric parameters of women submitted to vitamin D supplementation, even without regular physical exercise. The supplemented group obtained a 7% improvement in peak expiratory flow, similarly forced vital capacity, according to the data presented in **Figure 1** [42].

In a prospective, double-blind, randomized study, vitamin D supplementation in asthmatic children reduced the risk of exacerbation triggered by acute respiratory tract infection and significantly improved FEV1, [43]. Another prospective study in

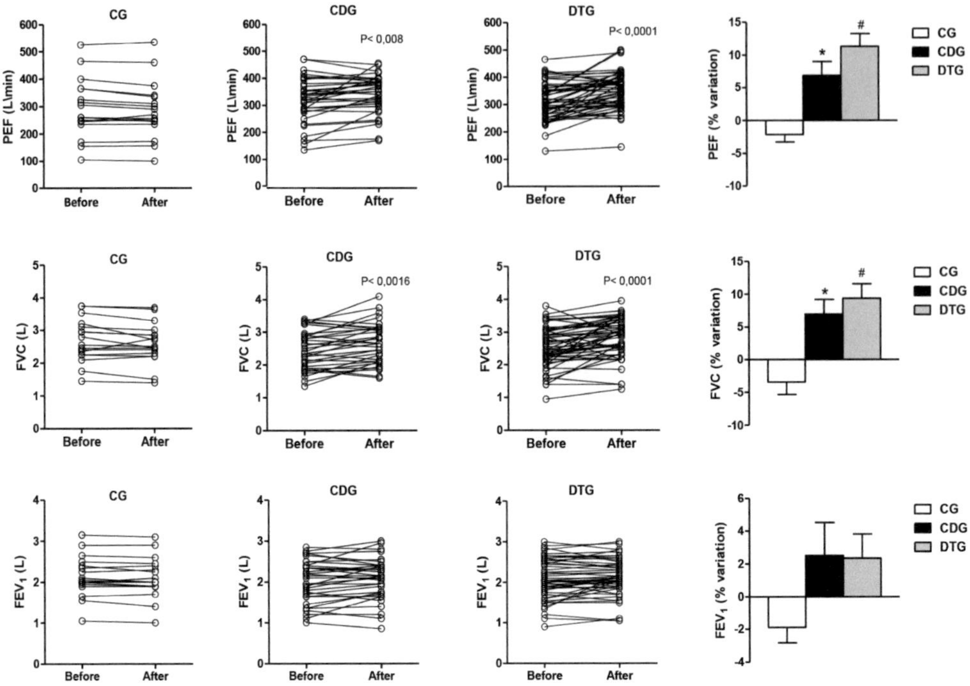

Figure 1.
*Follow-up of the spirometric parameters throughout the study of the three groups. Mean ± SD and percentage of change for peak expiratory flow (PEF), forced vital capacity (FVC) and forced expiratory volume in 1 s (FEV1) in the control (CG), control with supplementation of vitamin D and calcium (CDG), and training groups (DTG) before and after the intervention. * $p < 0.0001$, # $p < 0.01$ and $p < 0.05$ indicate differences detected using Student's t-test versus before intervention. Different letters indicate statistically significant differences ($p < 0.05$) on one-way analysis of variance (Tukey's post hoc test) in relation to the percentage variation between the groups. Ref. [42]: http://dx.doi.org/10.1590/2359-3997000000211.*

patients with chronic obstructive pulmonary disease (COPD) and vitamin D deficiency demonstrated increased respiratory muscle strength, improved dyspnea scale, and superior physical performance after 1 year of vitamin D supplementation [44].

In other components of respiratory system, vitamin D status has been associated with cardiorespiratory fitness in cross-sectional investigations in the general population. Regarding the maximum volume of oxygen (VO2max), in an observational study, no correlation was observed between 25 (OH) D levels and VO2max in 53 junior and collegiate ice hockey players. However, analysis of data soccer players exposure at ultraviolet radiation, revealed a linear association between vitamin D and VO2max in both experimental sessions [45]. Corroborating with these results, across from vitamin D supplementation of soccer players during an eight-week high-intensity training program, significant results were observed in aerobic capacity. As upshot, a significant improvement in VO2max in the supplemented group was observed compared to non-supplemented subjects [46].

A recent study found that more than 60% of athletes had vitamin D insufficiency even when the data collected near hot and sunny summer. Studies with high intensity athletes have been suggesting that athletes with vitamin D3 deficiency should be supplemented with this vitamin to improve physical performance, especially VO2max [47]. The explanation would be to athletes to achieve the best results in the summer, while exposure to solar radiation. The authors suggest that the replenished 25 (OH) D3 level may protect athletes against acute and chronic diseases. Considering the above, vitamin D3 supplementation along with the training load may induce adaptive changes of aerobic and anaerobic metabolism in athletes of different sports [47].

In rowers, significantly higher energy and oxygen consumption scores were observed during a continuous exercise test in the vitamin D3 supplemented group over the 8-week training period. They demonstrated a significantly increase in VO2max (12.1 and 10.3%, respectively) [21]. At the same time, the blood parameters of the supplemented athletes, such as IL-1b, CRP, LDH, reached lower values. These results suggest that vitamin D3, whose blood concentration increased by 400% in the supplemented group after supplementation, could be the justification for improving aerobic metabolism in rowers and reducing their inflammatory reactions in response to high intensity training [21].

The possible assumption by which vitamin D to affect VO2max would be caused the influence of the enzymes Cytochrome P450 (CYP) [48]. These enzymes activate vitamin D3, which has no hormonal action, converting it to an active hormonal form (1α, 25 (OH) 2D3) by the action of CYP enzymes. Reactions catalyzed by the CYP enzymes (mitochondrial CYP27A1, microsomal CYP2R1 in liver and the latter reaction by mitochondrial CYP27B1 in kidney) have proteins containing heme and could potentially affect the binding affinity of oxygen to hemoglobin [29, 49]. The existence of this compound is important for the transport of oxygen, since it is present mainly in hemoglobin, myoglobin and enzymes. Beside this, vitamin D could also influence VO2max through iron metabolism and erythropoietin. Complementing, the vitamin D deficiency results in dysregulation of innate immunity and inflammation which is affecting iron metabolism and contributes to erythropoietin resistance, and this well documented that is linearly associated with changes in red blood cells levels [47].

2.3 Vitamin D levels in athletes and physical performance

Even after 100 years its discovery, when early researchers suggested sunbathing to prevent and cure rickets, vitamin D remains in the spotlight. The actions of the vitamin D extend beyond bone health, becoming recognized at appropriate levels, beneficial to various non-skeletal health outcomes [49, 50]. Vitamin D is a multiactive hormone acting in different spheres of the body. Research over time has shown that vitamin D3 biological action is much broader than researchers originally thought, as shown by the tissue distribution of the VDR, from mediating only calcium homeostasis. Along with this, even after this recognition, it is visible from epidemiological data; that vitamin D deficiency excessively prevalent globally [50, 51].

Based on population data, the values of 25 (OH) D discussed in the literature with emphasis on bone outcomes range from 20 to 32 ng/mL (50–80 nmol/L). Several authors have confirmed that the best cut point of is 30 ng/mL (75 nmol/L) for the correction of secondary hyperparathyroidism, reduced risk of falls and fractures, and maximum absorption of calcium. Thus, serum concentrations below 20 ng/mL (50 nmol/L) are classified as deficiency, between 20 and 29 ng/mL (50 and 74 nmol/L) as insufficiency and between 30 and 100 ng/mL (75 and 250 ng/mL) as sufficiency [14, 52].

Surprisingly, it is estimated that around the world, 1 billion people fall into these categories. In addition, both vitamin D insufficiency and deficiency are increasing in prevalence [53]. Among athletes of different categories, deficiencies or deficiencies were observed in most dancers, taekwondo fighters, jockeys, elite wheelchair athletes, handball players, athletics athletes, weightlifters, swimmers and volleyball players. In relation to other professional sports, the athletes are affected in the same way. National Football League players, 26% were found to have deficient levels of vitamin D, and 42–80% of athletes had levels defined as insufficient. Similarly, professional basketball players, 32% of athletes are vitamin D deficient and 47% are insufficient [53].

The detection of vitamin D levels in athletes, in addition to the main function of vitamin D, acting in the interrelationship between the bone and muscular systems, also had as objective the sporting performance. Athletes need to potentiate the training stimulus, so it is a fundamental principle of the training program. The great scope of high performance training is to provide a stimulus to bring an adaptive response to the entire structure of the body that improves the performance of the competition. Thus, considering these aspects, the ergogenic resources to complement the adaptive response to a physical/metabolic challenge are intensively researched [53].

In the last decade, in vitro and animal studies have provided information on a beneficial role derived from vitamin D in skeletal muscle repair and remodeling. Although in humans the action of vitamin D within muscle tissue, still raise questions, in rodent were observed possible effects. The cytochrome precursor (CYP27B1) responsible for rendering vitamin D of 25 (OH) D3 inactive, in metabolically active (1,25 (OH) 2 D3), was found in cells at different stages of differentiation; expression in rodent muscle fibers [29, 54]. This finding suggests the action of vitamin D on the regulation of muscle tissue. In addition, recent studies suggest that VDR is expressed in myoblasts and C2C12 myotubes in murine skeletal muscle [54].

In human skeletal, in situ detection of VDR points towards a role of vitamin-D on muscle function. In addition, VDR has been localized to skeletal muscle cells that promote *de novo* protein synthesis. Considering the genomic effects, the activation of VDR induces heterodimerization between the active VDR and the retinoic receptor (RXR). In this way, this induces the activation of the vitamin D response element (VDRE), a complex of genes coding for the "genomic effects" of vitamin D [29, 54].

Among the genomic repercussions of VDR (**Figure 2**) [29], the increase in calcium handling by enhancing the activities of the calcium binding protein (calbindin-D9K) in cell sarcoplasm, muscle cell differentiation and proliferation through effects on insulin growth factor (IGF) expression which in turn induces skeletal muscle hypertrophy [29].

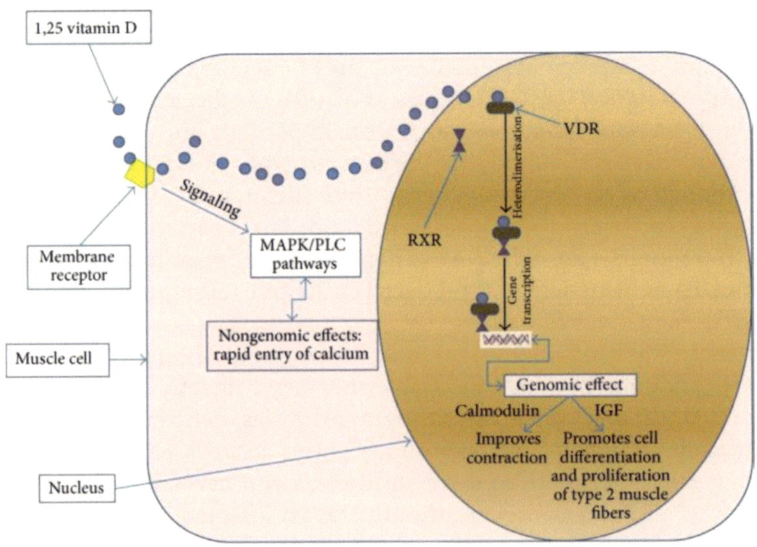

Figure 2.
Repercussions of the 1.25 vitamin D on skeletal muscle system: Molecular and nuclear pathways. Genomic and non-genomic effects. Ref. [29]: http://dx.doi.org/10.1155/2015/953241.

In an intervention study with 61 male athletes, it was observed that 62% had a serum concentration of 25 (OH) D of 20 ng/L at baseline and after supplementation with (5000 IU) of vitamin D3 per day for 8 weeks increased significantly 25 (OH) D concentrations and reflected on the velocity, verified through the 10-meter sprint time and the explosive force, through the vertical jump when compared to a placebo group [55].

Also in relation to physical performance, in a randomized, double-blind study, judo athletes supplemented with vitamin D3 achieved a 13% increase in muscle strength compared to a placebo group (p = 0.01) [56]. In a British study, through jump mechanography, was observed a positive association between serum vitamin D levels and jump height, velocity and power (p = 0.005, 0.002 and 0.003, respectively) in postmenarche adolescent girls [57].

Although some studies have demonstrated a possible outcome on physical performance [58], on the other hand, researchers investigated whether weekly supplementation with vitamin D3 at doses of 20,000 IU (500 IU) or 40,000 IU (1000 IU) for 12 weeks improved 25 (OH) D levels or performance measures in 30 club-level athletes. No correlations were observed between 25 (OH) D concentration and performance measures, indicating either vitamin D3 supplementation does not influence skeletal muscle function to induce measurable effect. These results were obtained even after 25 (OH) D levels were increased at the end of the study [59].

3. Conclusion

Vitamin D deficiency results in poor muscle function and weakness that are reversible upon reaching a complete vitamin D state. However, prospective studies examining the effects of vitamin D on muscle function have shown conflicting results, measures of physical performance of the distinct population (non-athletes, athletes, young and old). In this way, to make emphatic conclusions on the ergogenic use of vitamin D, we still have reflections on the potential mechanisms of interaction, or even, what would be the appropriate levels for the different age groups.

The direct mechanisms by which vitamin D could affect lung function have not yet been fully elucidated. Perhaps in future well-controlled studies, the parameters of cause and effect may be better supported. However, studies have shown that treatment with Vitamin D3 was effective in postmenopausal women to produce a significant increase in plasma concentrations of 25 (OH) D [42]. This adequacy of vitamin D status was associated with improved pulmonary function parameters, independent of the performance of the aquatic exercise program. In addition, the correction of its deficiency could also be a supporting measure for the strategies used to prevent and treat diseases with impaired pulmonary function or incapacitated individuals.

Furthermore, new questions are raised as to whether athletes amatours and professional susceptible to muscle damage and/or Vitamin D inadequacy, such as have been described in the elderly to exhibit low serum 25 [OH] D, experience aggravated declines in regenerative capacity and remodeling when serum 25 [OH] D is low. In addition to the health benefits, future studies may establish new views on the action of vitamin D as a possible legal ergogenic resource contributing to better athletic performance and record breaking.

Author details

Rodrigo Nolasco* and Marise Lazaretti-Castro
Division of Endocrinology, School of Medicine, Federal University of São Paulo/UNIFESP, São Paulo, Brazil

*Address all correspondence to: rodrigonolasco@yahoo.com.br

References

[1] Li Y, Pan A, Wang DD, Liu X, Dhana K, Franco OH, et al. Impact of healthy lifestyle factors on life expectancies in the US population. Circulation. 2018;**138**(4):345-355

[2] Wacker M, Holick MF. Vitamin D—Effects on skeletal and extraskeletal health and the need for supplementation. Nutrients. 2013;**5**(1):111-148

[3] Hossein-nezhad A, Holick MF. Vitamin D for health: A global perspective. Mayo Clinic Proceedings. 2013;**88**(7):720-755. DOI: 10.1016/j.mayocp.2013.05.011

[4] Nair R, Maseeh A. Vitamin D: The "sunshine" vitamin. Journal of Pharmacology and Pharmacotherapeutics. 2012;**3**(2):118-126

[5] Bikle DD. Vitamin D metabolism, mechanism of action, and clinical applications. Chemistry & Biology. 2014;**21**:319-329

[6] Dahlquist DT, Dieter BP, Koehle MS. Plausible ergogenic effects of vitamin D on athletic performance and recovery. Journal of the International Society of Sports Nutrition. 2015;**12**:33

[7] Dusso AS, Brown AJ, Slatopolsky E. Vitamin D. American Journal of Physiology. Renal Physiology. 2005;**289**:F8-F28

[8] Holick MF. Vitamin D: Evolutionary, physiological and health perspectives. Current Drug Targets. 2011;**12**(1):4-18

[9] Sutton AL, PN MD. Vitamin D: More than a "bone-a-fide" hormone. Molecular Endocrinology. 2003;**17**:777-791

[10] Gröber U, Spitz J, Reichrath J, Kisters K, Holick MF. Vitamin D: Update 2013: From rickets prophylaxis to general preventive healthcare. Dermato-Endocrinology. 2013;**5**(3):331-347

[11] Wang T-T, Tavera-Mendoza LE, Laperriere D, Libby E, MacLeod NB, Nagai Y, et al. Large-scale in silico and microarray-based identification of direct 1,25-dihydroxyvitamin D3 target genes. Molecular Endocrinology. 2005;**19**:2685-2695

[12] Ogan D, Pritchett K. Vitamin D and the athlete: Risks, recommendations, and benefits. Nutrients. 2013;**5**: 1856-1868

[13] Holick MF. Vitamin D deficiency. The New England Journal of Medicine. 2007;**357**(3):266-281

[14] Maeda SS, Borba VZ, Camargo MB, Silva DM, Borges JL, Bandeira F, et al. Brazilian Society of Endocrinology and Metabology (SBEM). Recommendations of the Brazilian Society of Endocrinology and Metabology (SBEM) for the diagnosis and treatment of hypovitaminosis D. Arquivos Brasileiros de Endocrinologia e Metabologia. 2014;**58**(5):411-433

[15] Bikle DD, Bouillon R. Vitamin D and bone and beyond. Bone Reports. 2018;**9**:120-121

[16] Barr R, Macdonald H, Stewart A, McGuigan F, Rogers A, Eastell R, et al. Association between vitamin D receptor gene polymorphisms, falls, balance and muscle power: Results from two independent studies [APOSS and OPUS]. Osteoporosis International. 2010;**21**:457-466

[17] Ceglia L, Niramitmahapanya S, da Silva Morais M, et al. A randomized study on the effect of vitamin d3 supplementation on skeletalmuscle morphology and vitamin d receptor concentration in older women.

Journal of Clinical Endocrinology and Metabolism. 2013;**98**(12):E1927-E1935

[18] Sato Y, Iwamoto J, Kanoko T, Satoh K. Low-dose vitamin D prevents muscular atrophy and reduces falls and hip fractures in women after stroke: A randomized controlled trial. Cerebrovascular Diseases. 2005;**20**:187-192

[19] Ryan JW, Anderson PH, Morris HA. Pleiotropic activities of vitamin D receptors–adequate activation for multiple health outcomes. Clinical Biochemist Reviews. 2015;**36**(2):53-61

[20] Norman AW. From vitamin D to hormone D: Fundamentals of the vitamin D endocrine system essential for good health. The American Journal of Clinical Nutrition. 2008;**88**:491S-499S. DOI: 10.1093/ajcn/88.2.491S

[21] Fitzgerald J, Peterson B, Warpeha J, Wilson P, Rhodes G, Ingraham S. Vitamin D status and VO2 peak during a skate treadmill graded exercise test in competitive ice hockey players. Journal of Strength and Conditioning Research. 2014;**28**:3200-3205

[22] Cangussu LM et al. Effect of isolated vitamin D supplementation on the rate of falls and postural balance in postmenopausal women fallers: A randomized, double-blind, placebo-controlled trial. Menopause. 2016;**23**(3):267-274

[23] Zhu K, Austin N, Devine A, Bruce D, Prince RL. A randomized controlled trial of the effects of vitamin D on muscle strength and mobility in older women with vitamin D insufficiency. Journal of the American Geriatrics Society. 2010;**58**(11):2063-2068. DOI: 10.1111/j.1532-5415.2010.03142.x

[24] Bischoff-Ferrari HA, Giovannucci E, Willett WC, Dietrich T. Dawson-Hughes B. Estimation of optimal serum concentrations of 25-hydroxyvitamin D for multiple health outcomes. The American Journal of Clinical Nutrition. 2006;**84**(1):18-28

[25] Moreira-Pfrimer LD, Pedrosa MA, Teixeira L, Lazaretti-Castro M. Treatment of vitamin D deficiency increases lower limb muscle strength in institutionalized older people independently of regular physical activity: A randomized double-blind controlled trial. Annals of Nutrition & Metabolism. 2009;**54**:291-300

[26] Moreira L, Fronza FC, Dos Santos RN, Teixeira LR, Kruel LF, Castro ML. High-intensity aquatic exercises (HydrOS) improve physical function and reduce falls among postmenopausal women. Menopause. 2013;**20**(10):1012-1019

[27] Wyon MA, Koutedakis Y, Wolman R, Nevill AM, Allen N. The influence of winter vitamin D supplementation on muscle function and injury occurrence in elite ballet dancers: A controlled study. Journal of Science and Medicine in Sport. 2014;**17**:8-12

[28] El-Hajj Fuleihan G, Nabulsi M, Tamim H, Maalouf J, Salamoun M, Khalife H, et al. Effect of vitamin D replacement on musculoskeletal parameters in school children: A randomized controlled trial. The Journal of Clinical Endocrinology and Metabolism. 2006;**91**:405-412

[29] Halfon M, Phan O, Teta D. Vitamin D: A review on its effects on muscle strength, the risk of fall, and frailty. BioMed Research International. 2015;**2015**:953241

[30] Uitterlinden AG, Fang Y, Van Meurs JB, Pols HA, Van Leeuwen JP. Genetics and biology of vitamin D receptor polymorphisms. Gene. 2004;**338**:143-156

[31] Banerjee A, Damera G, Bhandare R, Gu S, Lopez-Boado Y, Panettieri R

Jr, et al. Vitamin D and glucocorticoids differentially modulate chemokine expression in human airway smooth muscle cells. British Journal of Pharmacology. 2008;**155**(1):84-92

[32] Song Y, Qi H, Wu C. Effect of 1,25-(OH)2D3 (a vitamin D analogue) on passively sensitized human airway smooth muscle cells. Respirology. 2007;**12**:486-494

[33] Phokela SS, Peleg S, Moya FR, Alcorn JL. Regulation of human pulmonary surfactant protein gene expression by 1 alpha, 25-dihydroxyvitamin D3. American Journal of Physiology. Lung Cellular and Molecular Physiology. 2005;**289**(4):617-626

[34] Rehan VK, Torday JS, Peleg S, Gennaro L, Vouros P, Padbury J, et al. 1Alpha, 25-dihydroxy-3-epi-vitamin D3, a natural metabolite of 1alpha, 25-dihydroxy vitamin D3: Production and biological activity studies in pulmonary alveolar type II cells. Molecular Genetics and Metabolism. 2002;**76**:46-56

[35] Hansdottir S, Monick MM, Hinde SL, Lovan N, Look DC, Hunninghake GW. Respiratory epithelial cells convert inactive vitamin D to its active form: Potential effects on host defense. Journal of Immunology. 2008;**181**:7090-7099

[36] Nguyen M, Trubert CL, Rizk-Rabin M, et al. 1,25- Dihydroxyvitamin D3 and fetal lung maturation: Immunogold detection of VDR expression in pneumocytes type II cells and effect on fructose 1,6 bisphosphatase. The Journal of Steroid Biochemistry and Molecular Biology. 2004;**89-90**(1-5):93-97

[37] Nguyen TM, Guillozo H, Marin L, Tordet C, Koite S, Garabedian M. Evidence for a vitamin D paracrine system regulating maturation of developing rat lung epithelium. The American Journal of Physiology. 1996;**271**(3 Pt 1):L392-L399

[38] Gaultier C, Harf A, Balmain N, Cuisinier-Gleizes P, Mathieu H. Lung mechanics in rachitic rats. The American Review of Respiratory Disease. 1984;**130**(6):1108-1110

[39] Zosky GR, Berry LJ, Elliot JG, James AL, Gorman S, Hart PH. Vitamin D deficiency causes deficits in lung function and alters lung structure. American Journal of Respiratory and Critical Care Medicine. 2011;**183**:1336-1343

[40] Brun P, Dupret JM, Perret C, Thomasset M, Mathieu H. Vitamin D-dependent calcium-binding proteins (CaBPs) in human fetuses: Comparative distribution of 9K CaBP mRNA and 28K CaBP during development. Pediatric Research. 1987;**21**(4):362-367

[41] Black PN, Scragg R. Relationship between serum 25- hydroxyvitamin d and pulmonary function in the third national health and nutrition examination survey. Chest. 2005;**128**:3792-3798

[42] Nolasco R, Moreira LD, Bocalini DS, Fronza FC, Marin RV. Lazaretti-Castro M. Effects of vitamin D supplementation on pulmonary function in postmenopausal women following an aquatic exercise program. Archives of Endocrinology and Metabolism. 2017;**61**(1):28-35

[43] Majak P, Olszowiec-Chlebna M, Smejda K. Stelmach I vitamin D supplementation in children may prevent asthma exacerbation triggered by acute respiratory infection. The Journal of Allergy and Clinical Immunology. 2011;**127**:1294-1296

[44] Rezk NASA, Aly NYA, Hewidy AAH. Effect of vitamin D replacement in chronic obstructive pulmonary disease patients with vitamin D deficiency. The Egyptian Journal of Chest Diseases and Tuberculosis. 2015;**64**(2):353-357

[45] Koundourakis NE, Androulakis NE, Malliaraki N, Margioris AN. Vitamin D and exercise performance in professional soccer players. PLoS One. 2014;**9**(7):e101659. DOI: 10.1371/journal.pone.0101659

[46] Jastrzębska M et al. Can supplementation of vitamin d improve aerobic capacity in well trained youth soccer players? Journal of Human Kinetics. 2018;**61**:63-72

[47] Cannell JJ, Hollis BW, Sorenson MB, Taft TN, Anderson JJB. Athletic performance and vitamin D. Medicine and Science in Sports and Exercise. 2009;**41**:1102-1110

[48] Sugimoto H, Shiro Y. Diversity and substrate specificity in the structures of steroidogenic cytochrome P450 enzymes. Biological & Pharmaceutical Bulletin. 2012;**35**:818-823

[49] Holick MF. High prevalence of vitamin D inadequacy and implications for health. Mayo Clinic Proceedings. 2006;**81**(3):353-373. DOI: 10.4065/81.3.353

[50] Forrest KY, Stuhldreher WL. Prevalence and correlates of vitamin D deficiency in US adults. Nutrition Research. 2011;**31**(1):48-54. DOI: 10.1016/j.nutres.2010.12.001

[51] Ginde AA, Liu MC, Camargo CA Jr. Demographic differences and trends of vitamin D insufficiency in the US population, 1988- 2004. Archives of Internal Medicine. 2009;**169**(6):626-632. DOI: 10.1001/archinternmed.2008.604

[52] Sikora-Klak J et al. The effect of abnormal vitamin d levels in athletes. The Permanente Journal. 2018;**22**:17-216

[53] Owens DJ, Allison R, Close GL. Vitamin D and the athlete: Current perspectives and new challenges. Sports Medicine. 2018;**48**(Suppl 1):3-16. DOI: 10.1007/s40279-017-0841-9

[54] Olsson K, Saini A, Stromberg A, et al. Evidence for vitamin D receptor expression and direct effects of 1 alpha, 25(OH)2D3 in human skeletal muscle precursor cells. Endocrinology. 2016;**157**:98-111

[55] Close GL, Russell J, Cobley JN, et al. Assessment of vitamin D concentration in non-supplemented professional athletes and healthy adults during the winter months in the UK: Implications for skeletal muscle function. Journal of Sports Sciences. 2013;**31**(4):344-353

[56] Wyon MA, Wolman R, Nevill AM, et al. Acute effects of vitamin D3 supplementation on muscle strength in judoka athletes: A randomized placebo-controlled, double-blind trial. Clinical Journal of Sport Medicine. 2016;**26**(4):279-284. DOI: 10.1097/jsm.0000000000000264

[57] Ward KA, Das G, Berry JL, et al. Vitamin D status and muscle function in post-menarchal adolescent girls. The Journal of Clinical Endocrinology and Metabolism. 2009;**94**(2):559-563. DOI: 10.1210/jc.2008-1284

[58] Stockton KA, Mengersen K, Paratz JD, et al. Effect of vitamin D supplementation on muscle strength: A systematic review and meta-analysis. Osteoporosis International. 2011;**22**:859-871

[59] Close GL, Leckey J, Patterson M, et al. The effects of vitamin D(3) supplementation on serum total 25[OH]D concentration and physical performance: A randomised dose-response study. British Journal of Sports Medicine. 2013;**47**(11):692-696

Chapter 4

Behavioral Effects of Vitamin D_3 at Estrogen Deficiency in Females of Different Age

Julia Fedotova

Abstract

The present study was performed to determine the behavioral effects of cholecalciferol treatment at different doses as an adjunctive therapy alone or in a combination with low dose of 17β-estradiol on anxiety-like behavior of the middle-aged (12–14 months) and old (16–18 months) female rats at 12 weeks after ovariectomy. Vitamin D_3 supplementation individually (as cholecalciferol at doses of 1.0, 2.5 or 5.0 mg/kg/day, s.c.) or in co-administration with 17β-estradiol (17β-E_2, 0.5 μg/ rat, s.c.) were given to the ovariectomized (OVX) rats of different age at 12 weeks after ovariectomy. Anxiety-related state was tested in the elevated plus maze (EPM) and light-dark test (LDT), as well behavioral reactivity was registered in the open field test (OFT). The results of the study indicated that vitamin D_3 supplementation at doses of 1.0 or 5.0 mg/kg/day decreased manifestations of anxiety-like profile in the middle-aged or old OVX rats, respectively. Vitamin D_3 (1.0 mg/kg/day) plus 17β-E_2 resulted in more profound anxiolytic-like effects in the old OVX rats than effects of both drugs administered alone. Cholecalciferol (1.0 mg/kg/day, s.c.) in the old OVX rats produced elevated estradiol and 25-OH-VD_3 levels for these rats as compared to the old OVX females treated with solvent.

Keywords: vitamin D_3, anxiety, menopause, estradiol, aging, female rats

1. Introduction

Anxiety disorders are twice common in women than in men and the risk increased more during the menopausal transition [1, 2]. Menopause indicates the termination of a woman's reproductive life. It is defined as the permanent cessation of menstruation induced by the loss of ovarian follicular activity [2, 3]. Menopause represents an important stage in female lives, which is often associated with a plethora of complaints and sufferings. Mood disturbances, especially anxiety and depression, are commonly associated with menopause [3, 4]. For women, data suggest that estrogens are strongly implicated in the regulation of mood and behavior, as well as in the pathophysiology of mood disorders [5–7]. Nowadays, there is a strong tendency to increase the duration of living in the whole world, however, the median age of programmed termination of a woman's reproductive life failed to alter [5–7]. Thus, majority of women in the aging population lives significant period of their lifetime in a postmenopausal state which is clearly associated with very low

 IntechOpen

estrogen levels that could be one of the marked trigger factors for development of affective-related disorders [5, 7].

A strategy to alleviate the mood disorders associated with menopause is hormonal replacement therapy (HRT) [8]. However, controversial results related to the effectiveness of such treatment have been frequently reported [9]. These discrepancies could be associated to various factors, one of them being the time when estrogen restitution is initiated after the beginning of menopause [10, 11].

There is a growing interest about the potential of diet and nutrients to improve the mental health of the women population and for the treatment of psychiatric disorders [10, 11]. In the case of mood disorders, the limitations of psychotropic drugs to achieve adequate rates of clinical remission and functional recovery have promoted the search for complementary approaches [12–15]. Menopausal women are now choosing to take alternate and complementary therapies marketed as "natural" treatments that offer the positive health effects of estrogens without the unwanted side effects [12, 13]. Among other nutraceuticals, one of such "natural" substances for treatment of affective-related diseases could be vitamin D (VD) [14, 15].

VD is a neuroactive secosteroid with well-known skeletal physiological role and diverse "non-skeletal" functions in the human body [16, 17]. "Non-skeletal" functions of VD are connected with its different range of outcomes in the central nervous system, such as neuroplasticity, apoptosis, cell proliferation and differentiation [18, 19]. All functions of VD in the body (classical functions, i.e., effect upon calcium-phosphate management and the non-classical ones) are imposed by the nuclear VD receptor (VDR), regulating directly the gene expression [20, 21]. Nuclear VDR are member of receptors family for transcription factors which are activated by numerous ligands [20, 21]. VDR are present in most tissues and cells in the body, and within the brain it shows some specificity to the prefrontal cortex, hippocampus, cingulate gyrus, thalamus, hypothalamus and substantia nigra [22]. This is of relevance as many of those brain regions have been implicated in the physiology of affective-related disorders.

Estrogen deficiency effects on affective-related behavior are restricted to certain periods of age after ovary removal [23, 24]. Preclinical data suggest that onset age of menopause can be important to obtain behavioral positive or negative results. Thus, it is of great interest to evaluate the effects of repeated cholecalciferol administration on anxiety-related behavior in the middle-aged and old female rats with long-term estrogen deficiency.

The aim of the present study was to determine if repeated systemic treatment with cholecalciferol affects anxiety-like behavior in the middle-aged and old female rats after long-term ovariectomy.

2. Materials and methods

2.1 Animals

Female albino Wistar rats (12–14 months, middle-aged rats or 16–18 months, old rats, weighing 230–240 or 260–270 g, respectively) from the special biocollection of Koltushi vivarium (St. Petersburg, Russia) were used in the present study. All rats were allocated in groups and were allowed to accommodate for 1 week in the animal house at I.P. Pavlov Institute of Physiology, of the Russian Academy of Sciences, before subjecting them to behavioral testing and pharmacological treatments. They were provided with a standard pellet diet and were given water ad libitum. The animals were kept at a temperature of 23 ± 2°C and a 12 h light/dark cycle as well as a constant relative humidity (50 ± 10%) during all experimental sessions. Female rats of different age were randomly separated into experimental groups accordingly to their age, including the control groups.

Vitamin D_3 and 17β-E_2 treatments, as well as anxiety-related tests were carried out by double-blind method by using rules of the Health guide for the care and use of Laboratory animals (1978) formulated by the National Institute of Health. Females of different age were placed in the special room for behavioral trials at least 1 h prior to the beginning of the experimental sessions which were performed from 09:00 am to 12:00 am. The experimental protocols of this study were approved by the Institutional Animal Ethics Committee of I.P. Pavlov Institute of Physiology, Russia (protocol 1095/1 from June 25, 2012).

2.2 Surgery

Long-term ovariectomy surgery was performed as previously described [25]. Briefly, middle-aged and old female rats were anesthetized with ketamine (70 mg/kg b.w.) mixed with xylazine (10 mg/kg b.w.). To avoid inflammation, the rats were administered with meloxicam (1 mg/kg b.w.). The fallopian tube was crushed and the ovary was removed by cutting. The effectiveness of long-term ovariectomy or 17β-estradiol (17β-E_2) application was assessed by vaginal smears. The ovariectomized (OVX) females of different age were housed in groups of five in cages separated by groups. To assure the long-term absence of estrogens, all rats after surgery were remained to the housing facilities for 12 weeks.

2.3 Drug treatments

17β-estradiol, 17β-E_2 (Sigma, USA) at low dose of 5.0 μg/rat [7, 26] and vitamin D_3 as cholecalcirefol (Sigma, USA) at several doses (1.0, 2.5 or 5.0 mg/kg) [27] were subcutaneously (s.c.) administered once daily starting 14 days prior to the cognitive experiments. 17β-E_2 was dissolved in sterile sesame oil, $VD_3$was dissolved in 95% ethanol solvent, aliquoted and stored at −80°C. The stock of VD_3 was dissolved in sterile water, resulting in a solution of cholecalciferol with 2% ethanol. All drug solutions were freshly prepared before each behavioral testing. 17β-E_2 and cholecalcirefol were injected in a volume of 0.1 ml. The middle-aged OVX females were 15.5–17.5 months middle-aged at the onset for drug treatments.

The estrogen, 17β-E_2 (E-8875, Sigma Chemical Co., St. Louis, MO, USA) was dissolved in sterile sesame oil. Cholecalcirefol (C-9756, Sigma Chemical Co., St. Louis, MO, USA) was dissolved in 95% ethanol, aliquoted and stored at −80°C. The stock of cholecalciferol was diluted in sterile water, resulting in a solution of cholecalciferol with 2% ethanol. 17β-E_2 was injected subcutaneously (s.c. at a dose of 0.5 μg/rat).
The low dose of 17β-E_2 (5.0 μg/rat subcutaneously, s.c.) was chosen from the studies performed by Estrada-Camarena and co-workers [26, 28]. Three doses of cholecalciferol (1.0, 2.5 or 5.0 mg/kg, s.c.) were chosen from the behavioral study performed by Idrus and co-workers [27]. All solutions were freshly prepared before each experimental series. All preparations were administered in a volume of 0.1 ml. Following 12 weeks after ovariectomy, cholecalciferol, 17β-E_2 and oil solvent were injected once daily for 14 days.

2.4 Animal groups

Female rats (middle-aged or old, intact and OVX) were randomly divided into 24 groups, accordingly to their age, with 8 rats in each group.

The following experimental groups for the middle-aged and old female rats were created in the present study:

1 and 2—middle-aged or old intact female rats + solvent.

3 and 4—middle-aged or old intact female rats + cholecalciferol 1.0 mg/kg (middle-aged or old intact – vitamin D_3 1.0).

5 and 6—middle-aged or old intact female rats + cholecalciferol 2.5 mg/kg (middle-aged or old intact – vitamin D_3 2.5).

7 and 8—middle-aged or old intact female rats + cholecalciferol 5.0 mg/kg (middle-aged or old intact – vitamin D_3 5.0).

9 and 10—middle-aged or old OVX + solvent (middle-aged or old OVX – Sol).

11 and 12—middle-aged or old OVX rats + 17β-E_2 (middle-aged or old OVX – 17β-E_2).

13 and 14—middle-aged or old OVX rats + cholecalciferol 1.0 mg/kg (middle-aged or old OVX – vitamin D_3 1.0).

15 and 16—middle-aged or old OVX rats + cholecalciferol 2.5 mg/kg (middle-aged or old OVX – vitamin D_3 2.5).

17 and 18—middle-aged or old OVX rats + vitamin D_3 5.0 mg/kg (middle-aged or old OVX – vitamin D_3 5.0).

19 and 20—middle-aged or old OVX rats + cholecalciferol 1.0 mg/kg + 17β-E_2 (middle-aged or old OVX – vitamin D_3 1.0 – 17β-E_2).

21 and 22—middle-aged or old OVX rats + cholecalciferol 2.5 mg/kg + 17β-E_2 (middle-aged or old OVX – vitamin D_3 2.5 – 17β-E_2).

23 and 24—middle-aged or old OVX rats + cholecalciferol 5.0 mg/kg + 17β-E_2 (middle-aged or old OVX – vitamin D_3 5.0 – 17β-E_2).

The treatment period for animals was 14 days, and at the end of the treatment period (1 h after the last dose of solvent, vitamin D_3 or 17β-E_2), all animals were subjected to the elevated plus maze (EPM), light-dark test (LDT) and the open field test. During testing sessions in all behavioral tests, the control and experimental groups of rats were also given with solvent, vitamin D_3 or 17β-E_2.

2.5 Behavioral tests

2.5.1 Elevated plus maze test

EPM is commonly accepted as standard test of anxiety-like behavior and was used to assess anxiety-like behavioral responses [29, 30]. This test is sensitive to putative anxiogenic-like and anxiolytic-like drugs [31]. EPM consist of two open arms (50 × 10 cm^2) and two closed arms (40 × 10 cm^2) with a central platform (10 × 10 cm^2) and elevated 50 cm above the floor level. All female rats from control and experimental groups were randomly placed at the center of the EPM and allowed them to freely move in the apparatus for 5 min. The number of entries and total time spent in open arms were accepted as parameters of anxiolytic-like effects of treatments. The apparatus was cleaned with damp cloth after each trial to avoid place preference and the influence of olfactory stimuli.

2.5.2 Light/dark test

The apparatus consist of two identical boxes (30 × 40 × 40 cm), one of which with white walls and floor and illuminated by a 60 Watt light from above, while the other of the box was painted black and had a lid so it was not illuminated [32, 33]. The number of entrance and the total time in the light box were registered for 5 min [7]. The increase in the number of entrances and the total time in the light box were postulated as manifestation of anxiolytic-like effects of treatments. The apparatus was cleaned with damp cloth after each trial to avoid place preference and the influence of olfactory stimuli.

2.5.3 Open field test

The effect of cholecalciferol on locomotor, rearing and grooming activities was evaluated automatically using an open-field computer-aided controlling system as described previously [34]. The apparatus consists of a square platform (80.0 cm × 80.0 cm; wall height 36.0 cm). The floor of the platform was divided into 16 equal squares of 19.5 cm × 19.5 cm. A video camera fixed at the top, and the apparatus was illuminated by a light source of 120 Lux on the ceiling. Each rat was placed at the center of the apparatus and allowed to explore freely for 5 min. Total number of central and peripheral square crossings were recorded for each animal. The apparatus was cleaned with damp cloth after each trial to avoid place preference and the influence of olfactory stimuli.

2.6 Determination of estradiol, 25-OH-VD_3 and calcium levels in the blood serum

Blood samples were collected in tubes and centrifuged. After centrifugation, serum was separated, frozen and stored at −20°C until biochemical assessment. Estradiol levels were assessed using commercial available ELISA kit (DRG Diagnostics, Marburg, Germany). The sensitivity of the estradiol ELISA kit was 1.0 pg/ml. Measurement of 25-hydroxyvitamin D_3 (25-OH-VD_3) levels was performed by ELISA kit (CSB-E08098r, Cusabio Biotech Co., Ltd., Wuhan, P.R. China). Technical variability for 25-OH-VD_3 ELISA kit was low with coefficients of variation of <10% intra-assay and < 15% inter-assay. Detection range of 25-OH-VD_3 levels was 20–100 μg/L. The sensitivity of the 25-OH-VD_3 using ELISA kit was 5.0 μg/L. Calcium concentrations were detected by spectrophotometric method using calcium assay colorimetric kit (ab102505, Abcam, France). The sensitivity of the calcium kit was 0.1 m/M. All the procedures of estradiol, 25-OH-VD_3 and calcium kits were conducted following the manufacturer's instruction manual.

2.7 Statistical analysis

Data were expressed as means ± standard error (SEM). Differences among means were postulated as significant at $p \leq 0.05$. Behavioral and biochemical data were analyzed using a two-way ANOVA and subsequent *post-hoc* analysis was conducted with Dunnett's multiple comparison test. Statistical calculation was carried out using SPSS software 19 version (SPSS Inc., Chicago, IL., USA).

3. Results

3.1 Vitamin D_3 in different doses decreases anxiety-like profile of the middle-aged and old OVX and OVX rats given with 17β-estradiol after long-term absence of estrogen as measured in the EPM test

For vitamin D_3 supplementation, two-way ANOVA analysis revealed a significant interaction between hormone condition and treatments ([$F(5,44) = 12.83$, $p < 0.05$] and [$F(5,44) = 9.47, p < 0.01$], respectively), with significant effects of hormone conditions ([$F(5,44) = 9.47, p < 0.01$ and [$F(5,44) = 7.88, p < 0.01$], respectively) and treatment ([$F(5,44) = 15.24, p < 0.05$] and [$F(5,44) = 11.02, p < 0.05$], respectively) in the time spent into the open arms or the number of entries into the open arms of the middle-aged OVX rats. The *post-hoc* test demonstrated significant differences for these groups ($p < 0.05$).

Vitamin D_3 application at all tested doses did not modify the time spent into the open arms and the number of entries into the open arms in the intact middle-aged rats as compared to the control rats (**Figure 1a, b,** $p > 0.05$). Following 12 weeks of post-surgery period, the middle-aged female rats showed a profound decrease of the time spent in the open arms and the number of entries into the open arms as compared to the control females (**Figure 1a, b,** $p < 0.05$). The 17β-E_2 injection (0.5 μg/kg, s.c.) resulted in an increase of the time spent in the open arms and the number of entries into the open arms in the middle-aged OVX rats as compared to the middle-aged OVX rats administered with solvent (**Figure 1a, b,** $p < 0.05$), but did not reach the values of control rats (**Figure 1a, b,** $p < 0.05$).

Vitamin D_3 treatment at dose of 5.0 mg/kg significantly increased the time spent in the open arms and the number of entries into the open arms of the middle-aged OVX rats as compared to the middle-aged OVX rats given with solvent (**Figure 1a, b,** $p > 0.05$). Co-administration of vitamin D_3 at dose of 5.0 mg/kg and 17β-E_2 to the middle-aged OVX rats produced a more greater increase of the time spent in the open arms and the number of entries into the open arms than that of the middle-aged OVX rats given 17β-E_2 or solvent (**Figure 1a, b,** $p > 0.05$). However, vitamin D_3 treatment at doses of 1.0 and 2.5 mg/kg individually or plus 17β-E_2 failed to alter the time spent in the open arms and the number of entries

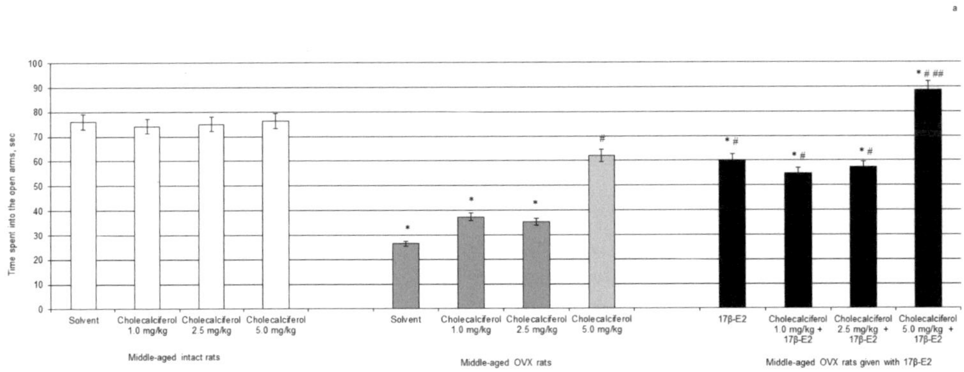

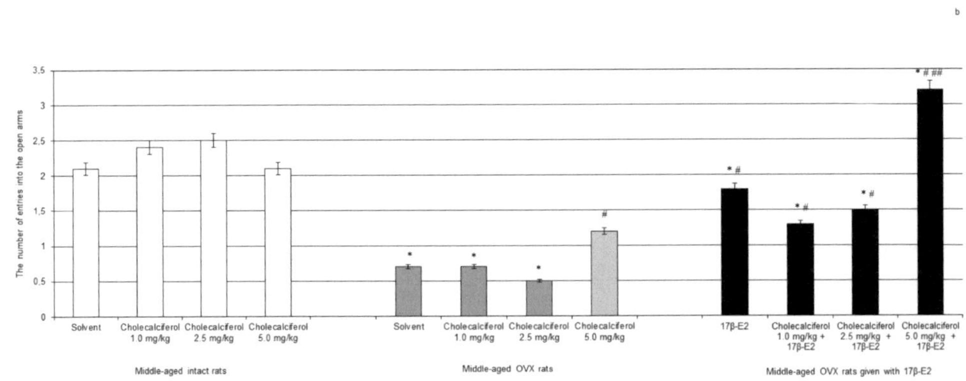

Figure 1.
*Effects of cholecalciferol administration on anxiety-like behavior of the middle-aged ovariectomized (OVX) rats following long-term estrogen deficiency in the elevated plus maze. (a) Time spent into the open arms, sec; (b) The number of entries into the open arms. The obtained results show the mean ± standard error of the mean (SEM). *—$p < 0.05$ as compared to the control group of the old sham-operated rats; #—$p < 0.05$ as compared to the old OVX rats treated with solvent; ##—$p < 0.05$ as compared to the old OVX rats treated with 17β-estradiol (17β-E_2). Each group comprised a minimum of eight rats. Cholecalciferol was given at 1.0, 2.5 or 5.0 mg/kg/day subcutaneously (s.c.), once daily, for 14 days. The administered dose of 17β-estradiol was 0.5 μg/rat s.c., once daily, for 14 days.*

into the open arms of the middle-aged OVX as compared to the middle-aged OVX rats treated with 17β-E_2 **Figure 1a, b**, $p > 0.05$). The time spent in the open arms and the number of entries into the open arms in the middle-aged OVX rats administered with vitamin D_3 at these doses in combination with 17β-E_2 were lower than that for middle-aged control rats and were higher than for middle-aged OVX rats (**Figure 1a, b**, $p > 0.05$).

Two-way ANOVA test demonstrated a significant hormone condition × treatment interaction ([$F(5,44) = 11.44$, $p < 0.05$] and [$F(5,44) = 11.12$, $p < 0.01$], respectively), significant effect of hormone conditions ([$F(5,44) = 9.22$, $p < 0.01$] and [$F(5,44) = 16.88$, $p < 0.01$], respectively) and significant effect for treatment ([$F(5,44) = 11.56$, $p < 0.05$], and [$F(5,44) = 12.56$, $p < 0.05$], respectively) for the time spent into the open arms or the number of entries into the open arms of the old OVX rats. *Post-hoc* analyses revealed differences among the groups for the anxiety-like state during experimental sessions ($p < 0.05$).

The old intact rats treated with vitamin D_3 at doses of 1.0, 2.5 or 5.0 mg/kg failed to alter the time spent into the open arms and the number of entries into the open arms as compared to the old control rats (**Figure 2a, b**, $p > 0.05$). The old OVX rats given with solvent displayed a significant decrease of the time spent into the open arms and the number of entries into the open arms as compared to the old control rats (**Figure 2a, b**, $p > 0.05$).

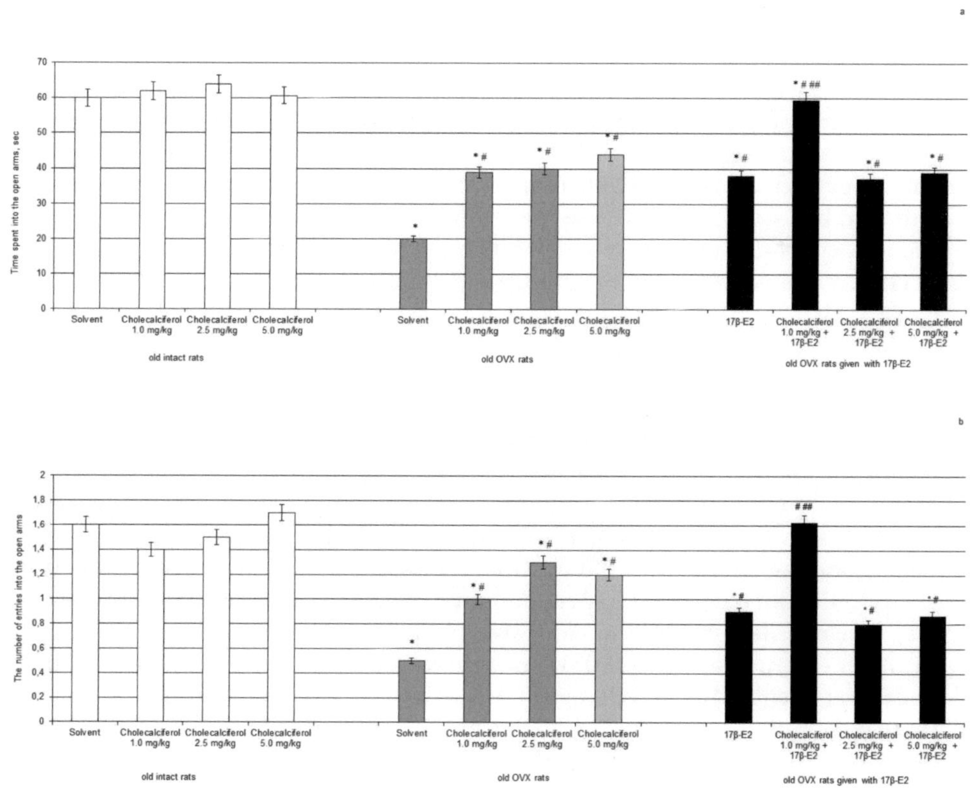

Figure 2.
*Effects of cholecalciferol administration on anxiety-like behavior of the old ovariectomized (OVX) rats following long-term estrogen deficiency in the elevated plus maze. (a) Time spent into the open arms, sec; (b) The number of entries into the open arms. The obtained results show the mean ± standard error of the mean (SEM). *—$p < 0.05$ as compared to the control group of the old sham-operated rats; #—$p < 0.05$ as compared to the old OVX rats treated with solvent; ##—$p < 0.05$ as compared to the old OVX rats treated with 17β-estradiol (17β-E_2). Each group comprised a minimum of eight rats. Cholecalciferol was given at 1.0, 2.5 or 5.0 mg/kg/day subcutaneously (s.c.), once daily, for 14 days. The administered dose of 17β-estradiol was 0.5 μg/rat s.c., once daily, for 14 days.*

Vitamin D_3 supplementation administered in all doses to the old OVX rats resulted in increase of the time spent into the open arms and the number of entries into the open arms as compared to the old OVX rats treated with solvent (**Figure 2a, b,** $p > 0.05$). Combined administration of cholecalciferol at dose of 1.0 mg/kg with 17β-E_2 to the old OVX rats more significantly increased the time spent into the open arms and the number of entries into the open arms as compared to the old OVX rats treated with solvent or 17β-E_2 (**Figure 2a, b,** $p > 0.05$). The old OVX rats administered with cholecalciferol at doses of 2.5 mg/kg and 5.0 mg/kg plus 17β-E_2 showed similar values of the time spent into the open arms and the number of entries into the open arms like as old OVX rats treated with 17β-E_2 (**Figure 2a, b,** $p > 0.05$).

3.2 Vitamin D_3 at different doses reverses anxiety-like profile of the middle-aged and old OVX and OVX rats given with 17β-estradiol after long-term absence of estrogen as measured in light: dark box

Two-way ANOVA statistical test revealed a significant interaction between hormone condition and treatments ([$F(5,44) = 11.52, p < 0.01$) and [$F(5,44) = 16.75, p < 0.05$], respectively), with significant effects of hormone conditions ([$F(5,44) = 19.22, p < 0.05$ and [$F(5,44) = 9.56, p < 0.001$], respectively) and treatment ([$F(5,44) = 9.88, p < 0.01$] and [$F(5,32) = 7.26, p < 0.05$], respectively) in the time spent and number of entries in the light compartment of the middle-aged OVX rats. The *post-hoc* test demonstrated significant differences for these groups ($p < 0.05$).

Vitamin D_3 supplementation at all doses did not alter the time spent and number of entries in the light compartment in the intact middle-aged rats as compared to the control rats (**Figure 3a, b,** $p > 0.05$). Following 12 weeks of post-ovariectomy period, the middle-aged female rats showed a profound decrease of the time spent and number of entries in the light compartment as compared to the control females (**Figure 3a, b,** $p < 0.05$). The 17β-E_2 injection (0.5 μg/kg, s.c.) resulted in an increase of the time spent and number of entries in the light compartment in the middle-aged OVX rats as compared to the middle-aged OVX rats administered with solvent (**Figure 3a, b,** $p < 0.05$), but did not reach the values of control rats (**Figure 3a, b,** $p < 0.05$).

Vitamin D_3 treatment at dose of 5.0 mg/kg markedly elevated time spent and number of entries in the light compartment of the middle-aged OVX rats as compared to the middle-aged OVX rats given with solvent (**Figure 3a, b,** $p > 0.05$). Co-administration of vitamin D_3 at dose of 5.0 mg/kg and 17β-E_2 to the middle-aged OVX rats produced greater increase of time spent and number of entries in the light compartment than that of the middle-aged OVX rats given 17β-E_2 or solvent (**Figure 3a, b,** $p > 0.05$). Vitamin D_3 treatment at doses of 1.0 mg/kg and 2.5 mg/kg individually or plus 17β-E_2 did not change time spent and number of entries in the light compartment of the middle-aged OVX as compared to the middle-aged OVX rats treated with 17β-E_2 (**Figure 3a, b,** $p > 0.05$). The time spent and number of entries in the light compartment in the middle-aged OVX rats administered vitamin D_3 at these doses in combination with 17β-E_2 were lower than that for middle-aged control rats and were higher than for middle-aged OVX rats (**Figure 3a, b,** $p > 0.05$).

Two-way analysis of variance showed a significant hormone condition × treatment interaction ([$F(5,44) = 11.52, p < 0.01$] and [$F(5,44) = 12.74, p < 0.05$], respectively), significant effect of hormone conditions ([$F(5,44) = 11.22, p < 0.05$] and [$F(5,44) = 11.56, p < 0.001$], respectively) and significant effect for treatment ([$F(5,44) = 10.08, p < 0.01$] and [$F(5,32) = 12.26, p < 0.05$)], respectively) for time spent and number of entries in the light compartment of the old OVX rats. *Post-hoc* analyses revealed differences among the groups for the anxiety-like state during experimental sessions ($p < 0.05$).

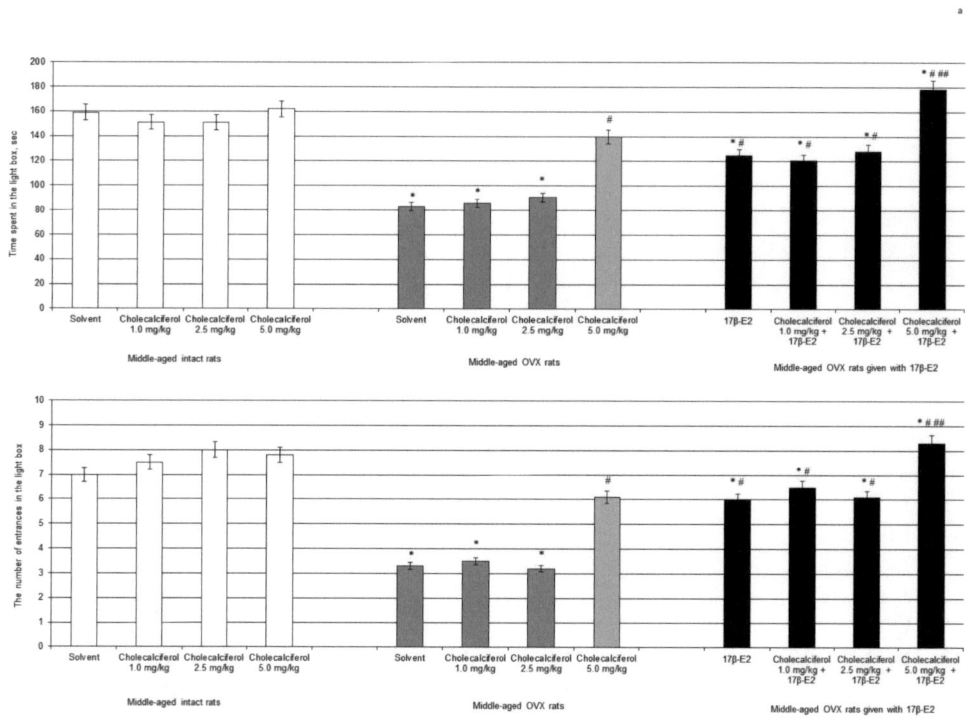

Figure 3.
*Effects of cholecalciferol administration on anxiety-like behavior of the middle-aged OVX rats following long-term estrogen deficiency in the light/dark test. (a) Time spent in the light box, sec; (b) The number of entrances in the light box. The obtained results show the mean ± standard error of the mean (SEM). *—$p < 0.05$ as compared to the control group of sham-operated rats, #—$p < 0.05$ as compared to the old OVX rats treated with solvent, ##—$p < 0.05$ as compared to the old OVX rats treated with 17β-estradiol. Each group comprised a minimum of eight rats. Cholecalciferol was given at 1.0, 2.5 or 5.0 mg/kg/day s.c., once daily, for 14 days. The administered dose of 17β-estradiol (17β-E_2) was 0.5 μg/rat s.c., once daily, for 14 days.*

The old intact rats treated with vitamin D_3 at doses of 1.0, 2.5 or 5.0 mg/kg failed to modify time spent and number of entries in the light compartment as compared to the old control rats (**Figure 4a, b**, $p > 0.05$). The old OVX rats given with solvent displayed a significant decrease of time spent and number of entries in the light compartment as compared to the old control rats (**Figure 4a, b**, $p > 0.05$).

Vitamin D_3 supplementation administered in all doses to the old OVX rats induced increase of the time spent and number of entries in the light compartment as compared to the old OVX rats treated with solvent (**Figure 4a, b**, $p > 0.05$). Combined administration of cholecalciferol at dose of 1.0 mg/kg with 17β-E_2 to the old OVX rats more significantly increased time spent and number of entries in the light compartment as compared to the old OVX rats treated with solvent or 17β-E_2 (**Figure 4a, b**, $p > 0.05$). The old OVX rats administered with vitamin D_3 at doses of 2.5 mg/kg and 5.0 mg/kg plus 17β-E_2 showed identical parameters of time spent and number of entries in the light compartment like as old OVX rats treated with 17β-E_2 (**Figure 4a, b**, $p > 0.05$).

3.3 Vitamin D_3 administration changes behavioral reactivity of the middle-aged and old OVX and OVX rats treated with 17β-estradiol

Accordingly to the two-way ANOVA test, there were significant differences for the grooming behavior between hormone conditions ([F(5,44) = 8.12, $p < 0.05$] and [F(5,44) = 9.22, $p < 0.05$], respectively) between drug treatment ([F(5,44) = 12.51,

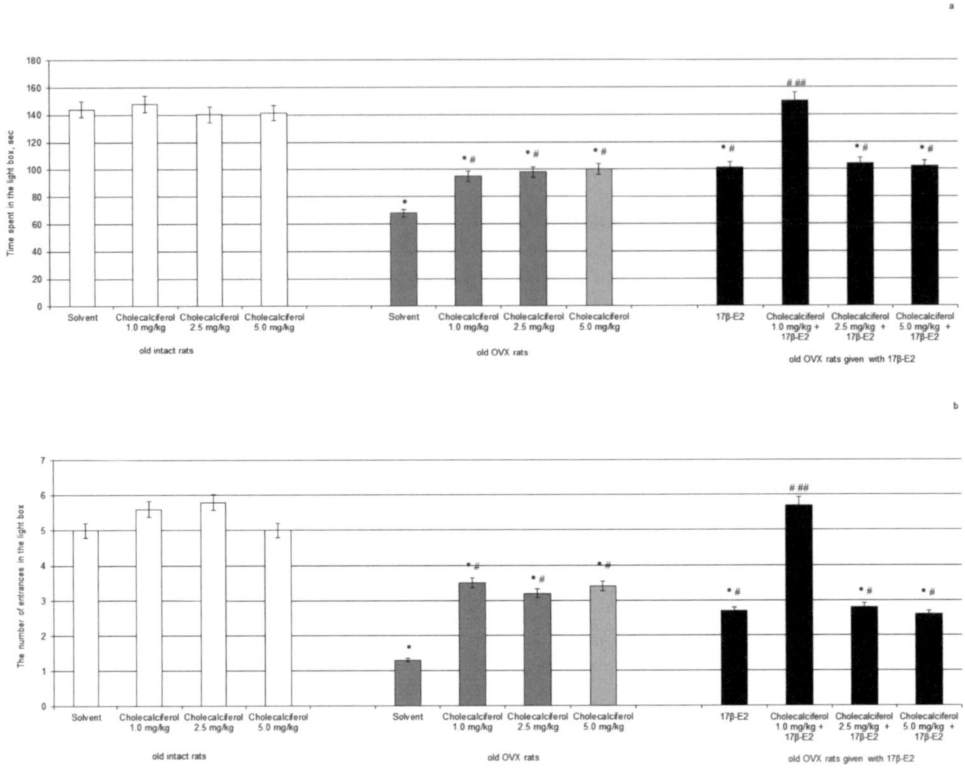

Figure 4.
*Effects of cholecalciferol administration on anxiety-like behavior of the old OVX rats following long-term estrogen deficiency in the light/dark test. (a) Time spent in the light box, sec; (b) The number of entrances in the light box. The obtained results show the mean ± standard error of the mean (SEM). *—$p < 0.05$ as compared to the control group of sham-operated rats, #—$p < 0.05$ as compared to the old OVX rats treated with solvent, ##—$p < 0.05$ as compared to the old OVX rats treated with 17β-estradiol. Each group comprised a minimum of eight rats. Cholecalciferol was given at 1.0, 2.5 or 5.0 mg/kg/day s.c., once daily, for 14 days. The administered dose of 17β-estradiol (17β-E_2) was 0.5 μg/rat s.c., once daily, for 14 days.*

$p < 0.01$] and [$F(5,44) = 12.56$, $p < 0.01$], respectively) and an interaction between hormone condition and treatments ([$F(5,44) = 7.16$, $p < 0.01$] and [$F(5,44) = 9.26$, $p < 0.01$], respectively) in the middle-aged and old OVX rats. Further *post-hoc* test revealed differences for grooming between experimental groups of the OVX rats with different age ($p < 0.05$).

Vitamin D_3 injected at several doses failed to demonstrate any changes of behavioral reactivity of the middle-aged and old intact females in the OFT as compared to the middle-aged control rats (**Tables 1** and **2**, $p > 0.05$).

A significant decrease of grooming behavior was registered in the middle-aged and old OVX rats given with solvent as compared to the control (**Tables 1** and **2**, $p < 0.05$). 17β-E_2 significantly reduced grooming reactions in the middle-aged and old OVX rats as compared to the middle-aged OVX rats (**Tables 1** and **2**, $p < 0.05$). The middle-aged and old OVX rats treated with vitamin D_3 in all tested doses alone or in a combination with 17β-E_2 did not demonstrate any modifications of motor and rearing activities as compared to the middle-aged OVX rats given with solvent (**Tables 1** and **2**, $p < 0.05$).

However, the middle-aged and old OVX rats treated with vitamin D_3 at doses of 1.0, 2.5 and 5.0 mg/kg demonstrated an increase of grooming behavior as compared to the middle-aged OVX rats. A co-administration of vitamin D_3 at these doses with 17β-E_2 decreased grooming behavior as compared to both middle-aged, as well as old intact and OVX rats received with solvent or 17β-E_2 (**Tables 1** and **2**, $p < 0.05$).

3.4 Modifications of 25-hydroxyvitamin D_3, estradiol and calcium levels in the blood serum following vitamin D_3 administration in the middle-aged and old OVX and OVX females treated with 17β-estradiol

The middle-aged intact rats treated with cholecalciferol at doses of 1.0, 2.5 and 5.0 mg/kg increased 25-OH-VD_3 levels (**Figure 5**, $p < 0.05$) and failed to alter estradiol levels in the serum blood as compared to the control rats (**Figure 6**, $p > 0.05$).

Long-term ovariectomy in the middle-aged female rats resulted in a significant decrease of estradiol and 25-OH-VD_3 levels in the blood as compared to the middle-aged control females (**Figures 5** and **6**, $p < 0.05$). The 17β-E_2 supplementation (0.5 μg/kg, SC) failed to modify 25-OH-VD_3 levels in the blood of the middle-aged OVX rats as compared to the middle-aged OVX rats administered with solvent (**Figure 5**, $p > 0.05$), and the value of this parameter in the middle-aged OVX/17β-E_2 females were lower than that of the value of middle-aged control rats. However, 17β-E_2 supplementation significantly increased estradiol levels in the blood of the middle-aged OVX rats as compared to the middle-aged OVX rats given with solvent (**Figure 6**, $p < 0.05$).

The middle-aged OVX rats treated with cholecalciferol at all tested doses significantly increased estradiol levels in the serum blood as compared to the middle-aged OVX rats treated with solvent (**Figure 6**, $p < 0.05$). However, the value of estradiol levels in the middle-aged OVX rats treated with cholecalciferol at these doses were lower than that of the value of middle-aged control rats. The middle-aged OVX rats treated with cholecalciferol at doses of 2.5 and 5.0 mg/kg significantly increased 25-OH-VD_3 levels in the serum blood as compared to the middle-aged OVX rats

Groups	Crossing	Rearing	Grooming
Middle-aged control rats + solvent	73.3 ± 4.2	12.1 ± 0.8	3.0 ± 0.2
Middle-aged intact rats + cholecalciferol 1.0 mg/kg	68.0 ± 2.4	12.0 ± .05	3.2 ± 0.2
Middle-aged intact rats + cholecalciferol 2.5 mg/kg	59.3 ± 5.6	10.5 ± 0.8	3.2 ± 0.2
Middle-aged intact rats + cholecalciferol 5.0 mg/kg	69.5 ± 4.2	11.7 ± 0.8	3.5 ± 0.2
Middle-aged OVX rats + solvent (OVX/solvent rats)	62.4 ± 2.3	12.3 ± 0.6	$1.2 \pm 0.5^{*}$
Middle-aged OVX rats +17β-E_2 (OVX/17β-E_2 rats)	60.3 ± 2.6	13.2 ± 0.3	$3.3 \pm 0.4^{\#}$
Middle-aged OVX rats + cholecalciferol 1.0 mg/kg	65.2 ± 2.5	12.1 ± 0.6	$4.0 \pm 0.2^{\#}$
Middle-aged OVX rats + cholecalciferol 2.5 mg/kg	76.9 ± 4.2	13.2 ± 0.8	$3.8 \pm 0.4^{\#}$
Middle-aged OVX rats + cholecalciferol 5.0 mg/kg	72.3 ± 4.4	11.0 ± 0.5	$4.1 \pm 0.6^{\#}$
Middle-aged OVX rats + cholecalciferol 1.0 mg/kg + 17β-E_2	62.1 ± 2.8	12.6 ± 0.6	$0.6 \pm 0.2^{*,\#,\#\#}$
Middle-aged OVX rats + cholecalciferol 2.5 mg/kg + 17β-E_2	73.2 ± 2.4	10.2 ± 0.4	$0.5 \pm 0.2^{*,\#,\#\#}$
Middle-aged OVX rats + cholecalciferol 5.0 mg/kg + 17β-E_2	65.4 ± 5.6	12.5 ± 0.8	$0.8 \pm 0.2^{*,\#,\#\#}$

$^{*}p < 0.05$ as compared to the control group of the old sham-operated rats.
$^{\#}p < 0.05$ as compared to the old OVX rats treated with solvent.
$^{\#\#}p < 0.05$ as compared to the old OVX rats treated with 17β-estradiol.
The obtained results show the mean ± S.E.M. Each group comprised a minimum of eight rats. Cholecalciferol was given at 1.0, 2.5 or 5.0 mg/kg/day, s.c., once daily, for 14 days. 17β-Estradiol (17β-E_2) was given at 0.5 μg/rat, s.c., once daily, during 14 days.

Table 1.
Cholecalciferol influences on behavioral parameters of the middle-aged OVX rats following long-term estrogen deficiency in the open field test for 5 min.

Groups	Crossing	Rearing	Grooming
Old control rats + solvent	69.7 ± 5.2	11.5 ± 0.3	2.9 ± 0.2
Old intact rats + cholecalciferol 1.0 mg/kg	71.8 ± 2.9	10.7 ± 0.3	3.2 ± 0.2
Old intact rats + cholecalciferol 2.5 mg/kg	66.9 ± 3.6	10.4 ± 0.2	3.1 ± 0.2
Old intact rats + cholecalciferol 5.0 mg/kg	69.0 ± 4.2	12.2 ± 0.8	3.0 ± 0.2
Old OVX rats + solvent (OVX/solvent rats)	72.1 ± 2.3	12.1 ± 0.6	$1.0 \pm 0.2^{*}$
Old OVX rats + 17β-E_2 (OVX/17β-E_2 rats)	64.3 ± 4.6	11.7 ± 0.8	$3.1 \pm 0.3^{\#}$
Old OVX rats + cholecalciferol 1.0 mg/kg	63.2 ± 3.5	12.6 ± 0.9	$4.2 \pm 0.2^{\#}$
Old OVX rats + cholecalciferol 2.5 mg/kg	67.2 ± 5.2	10.2 ± 0.8	$3.9 \pm 0.2^{\#}$
Old OVX rats + cholecalciferol 5.0 mg/kg	70.3 ± 4.4	11.5 ± 0.5	$4.3 \pm 0.2^{\#}$
Old OVX rats + cholecalciferol 1.0 mg/kg + 17β-E_2	72.1 ± 6.8	12.2 ± 0.6	$0.7 \pm 0.2^{*,\#,\#\#}$
Old OVX rats + cholecalciferol 2.5 mg/kg + 17β-E_2	78.5 ± 8.4	11.8 ± 0.4	$0.6 \pm 0.2^{*,\#,\#\#}$
Old OVX rats + cholecalciferol 5.0 mg/kg + 17β-E_2	69.4 ± 6.6	10.9 ± 0.8	$0.9 \pm 0.2^{*,\#,\#\#}$

$^{*}p < 0.05$ as compared to the control group of the old sham-operated rats.

$^{\#}p < 0.05$ as compared to the old OVX rats treated with solvent.

$^{\#\#}p < 0.05$ as compared to the old OVX rats treated with 17β-estradiol.

The obtained results show the mean ± S.E.M. Each group comprised a minimum of 8 rats. Cholecalciferol was given at 1.0, 2.5 or 5.0 mg/kg/day, s.c., once daily, for 14 days. 17β-Estradiol (17β-E_2) was given at 0.5 μg/rat, s.c., once daily, during 14 days.

Table 2.

Effects of cholecalciferol administration on behavioral impairments of the old OVX rats following long-term estrogen deficiency in the open field test for 5 min.

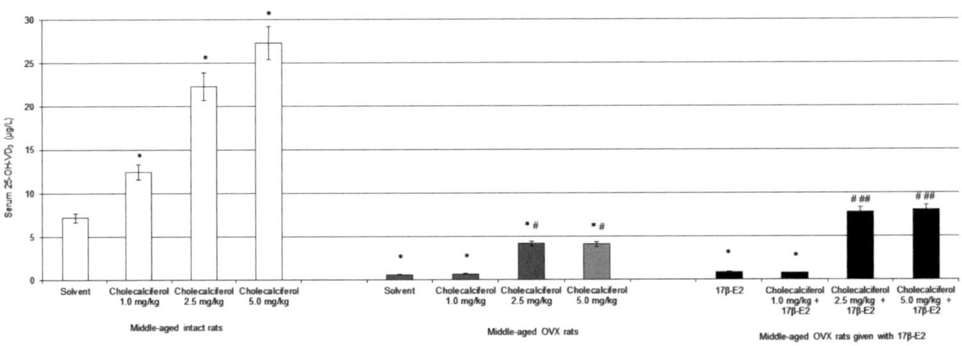

Figure 5.

*Effects of cholecalciferol administration on 25-OH-VD_3 level of the middle-aged ovariectomized (OVX) rats following long-term estrogen deficiency in the serum blood. The obtained results show the mean ± standard error of the mean (SEM). *—$p < 0.05$ as compared to the control group of the old sham-operated rats; #—$p < 0.05$ as compared to the old OVX rats treated with solvent; ##—$p < 0.05$ as compared to the old OVX rats treated with 17β-estradiol (17β-E_2). Each group comprised a minimum of eight rats. Cholecalciferol was given at 1.0, 2.5 or 5.0 mg/kg/day subcutaneously (s.c.), once daily, for 14 days. The administered dose of 17β-estradiol was 0.5 μg/rat s.c., once daily, for 14 days.*

treated with solvent (**Figure 5**, $p < 0.05$). The value of 25-OH-VD_3 content in the middle-aged OVX rats treated with cholecalciferol at doses of 2.5 and 5.0 mg/kg were lower than that of the value of middle-aged control rats. Moreover, cholecalciferol administered at a dose of 1.0 mg/kg into the middle-aged OVX rats failed to change 25-OH-VD_3 levels in the serum blood as compared to the middle-aged OVX rats treated with solvent (**Figure 5**, $p > 0.05$). Cholecalciferol treatment at doses of

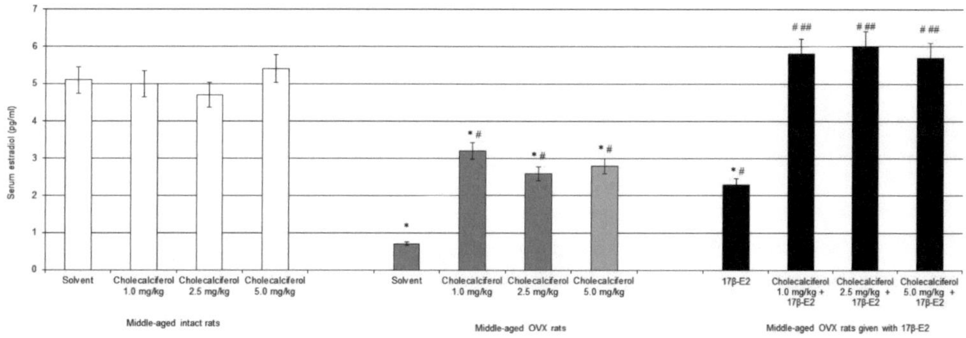

Figure 6.
*Effects of cholecalciferol administration on estradiol level of the middle-aged ovariectomized (OVX) rats following long-term estrogen deficiency in the serum blood. The obtained results show the mean ± standard error of the mean (SEM). *—$p < 0.05$ as compared to the control group of the old sham-operated rats; #—$p < 0.05$ as compared to the old OVX rats treated with solvent; ##—$p < 0.05$ as compared to the old OVX rats treated with 17β-estradiol (17β-E_2). Each group comprised a minimum of eight rats. Cholecalciferol was given at 1.0, 2.5 or 5.0 mg/kg/day subcutaneously (s.c.), once daily, for 14 days. The administered dose of 17β-estradiol was 0.5 μg/rat s.c., once daily, for 14 days.*

2.5 and 5.0 mg/kg in combination with 17β-E_2 more significantly elevated 25-OH-VD_3 levels for the middle-aged OVX rats as compared to the OVX females treated with oil solvent or 17β-E_2 (**Figure 3**, $p < 0.05$). Combined administration of cholecalciferol at a dose of 1.0 mg/kg and 17β-E_2 in the middle-aged OVX rats failed to change 25-OH-VD_3 levels as compared to the OVX rats administered with 17β-E_2 or solvent (**Figure 5**, $p > 0.05$). Cholecalciferol at all doses in combination with 17β-E_2 significantly increased estradiol levels when middle-aged OVX rats/cholecalciferol in tested doses plus 17β-E_2 rats were compared with the middle-aged OVX/solvent and OVX/17β-E_2 rat groups (**Figure 6**, $p > 0.05$).

The old intact rats treated with cholecalciferol at doses of 1.0, 2.5 and 5.0 mg/kg increased 25-OH-VD_3 levels (**Figure 7**, $p < 0.05$) and failed to alter estradiol levels in the serum blood as compared to the control rats (**Figure 8**, $p > 0.05$). Long-term ovariectomy in the old female rats resulted in a significant decrease of estradiol and 25-OH-VD_3 levels in the blood as compared to the old control females (**Figures 7** and **8**, $p < 0.05$). The 17β-E_2 supplementation (0.5 μg/kg, s.c.) failed to modify

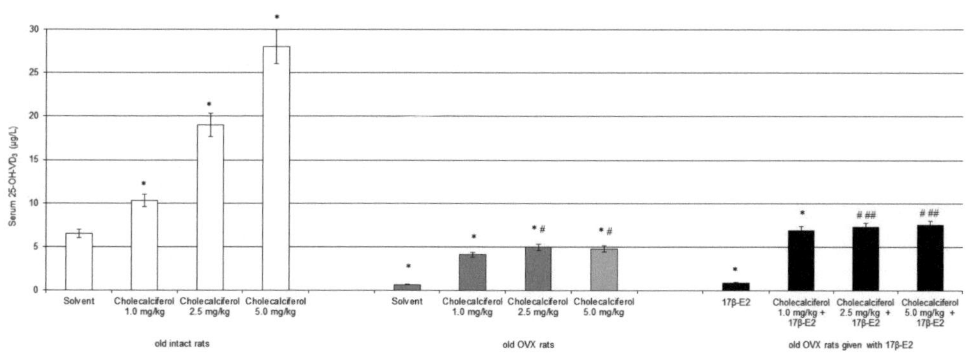

Figure 7.
*Effects of cholecalciferol administration on 25-OH-VD_3 level of the old ovariectomized (OVX) rats following long-term estrogen deficiency in the serum blood. The obtained results show the mean ± standard error of the mean (SEM). *—$p < 0.05$ as compared to the control group of the old sham-operated rats; #—$p < 0.05$ as compared to the old OVX rats treated with solvent; ##—$p < 0.05$ as compared to the old OVX rats treated with 17β-estradiol (17β-E_2). Each group comprised a minimum of eight rats. Cholecalciferol was given at 1.0, 2.5 or 5.0 mg/kg/day subcutaneously (s.c.), once daily, for 14 days. The administered dose of 17β-estradiol was 0.5 μg/rat s.c., once daily, for 14 days.*

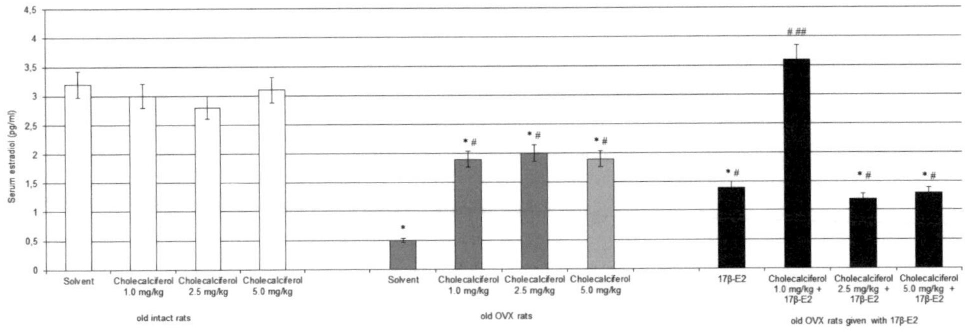

Figure 8.
*Effects of cholecalciferol administration on estradiol level of the old ovariectomized (OVX) rats following long-term estrogen deficiency in the serum blood. The obtained results show the mean ± standard error of the mean (SEM). *—$p < 0.05$ as compared to the control group of the old sham-operated rats; #—$p < 0.05$ as compared to the old OVX rats treated with solvent; ##—$p < 0.05$ as compared to the old OVX rats treated with 17β-estradiol (17β-E_2). Each group comprised a minimum of eight rats. Cholecalciferol was given at 1.0, 2.5 or 5.0 mg/kg/day subcutaneously (s.c.), once daily, for 14 days. The administered dose of 17β-estradiol was 0.5 μg/rat s.c., once daily, for 14 days.*

25-OH-VD_3 levels in the blood of the old OVX rats as compared to the old OVX rats administered with solvent (**Figure 7**, $p > 0.05$), and the value of this parameter in the old OVX/17β-E_2 females were lower than that of the value of old control rats. However, 17β-E_2 supplementation significantly increased estradiol levels in the blood of the old OVX rats as compared to the old OVX rats given with solvent (**Figure 8**, $p < 0.05$).

The old OVX rats treated with cholecalciferol at all doses significantly increased 25-OH-VD_3 and estradiol levels in the serum blood as compared to the old OVX rats treated with solvent (**Figure 8**, $p < 0.05$). However, the values of 25-OH-VD_3 and estradiol levels in the old OVX rats treated with cholecalciferol at all doses were lower than that of the values of old control rats.

Co-administration of vitamin D_3 (1.0 mg/kg) and 17β-E_2 markedly enhanced estradiol levels in the old OVX rats as compared to the groups of old OVX rats received with solvent or 17β-E_2 (**Figure 7**, $p < 0.05$). Vitamin D_3 supplementation (2.5 and 5.0 mg/kg) plus 17β-E_2 did not modify estradiol concentrations in the serum blood of the old OVX rats as compared to the OVX rats given with 17β-E_2

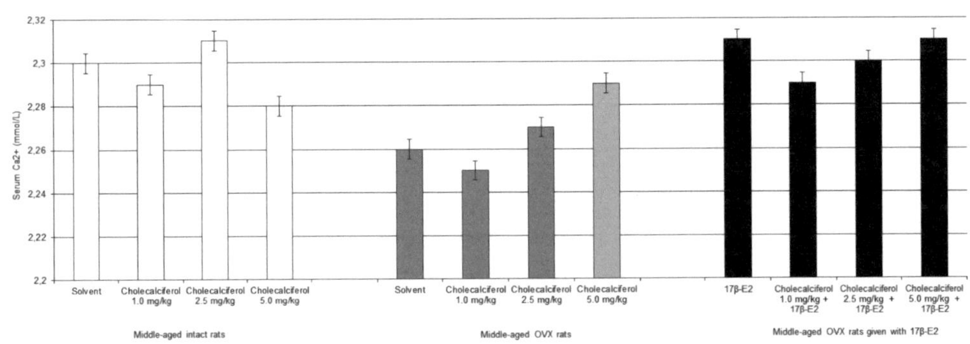

Figure 9.
*Effects of cholecalciferol administration on calcium level of the middle-ged ovariectomized (OVX) rats following long-term estrogen deficiency in the serum blood. The obtained results show the mean ± standard error of the mean (SEM). *—$p < 0.05$ as compared to the control group of the old sham-operated rats; #—$p < 0.05$ as compared to the old OVX rats treated with solvent; ##—$p < 0.05$ as compared to the old OVX rats treated with 17β-estradiol (17β-E_2). Each group comprised a minimum of eight rats. Cholecalciferol was given at 1.0, 2.5 or 5.0 mg/kg/day subcutaneously (s.c.), once daily, for 14 days. The administered dose of 17β-estradiol was 0.5 μg/rat s.c., once daily, for 14 days.*

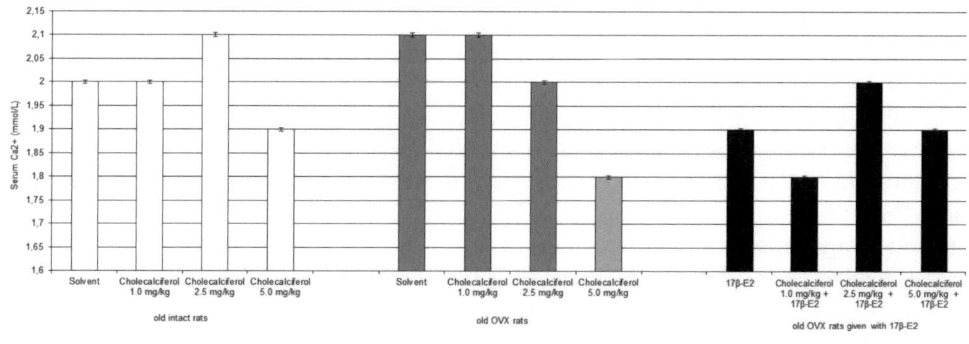

Figure 10.
*Effects of cholecalciferol administration on calcium level of the old ovariectomized (OVX) rats following long-term estrogen deficiency in the serum blood. The obtained results show the mean ± standard error of the mean (SEM). *—$p < 0.05$ as compared to the control group of the old sham-operated rats; #—$p < 0.05$ as compared to the old OVX rats treated with solvent; ##—$p < 0.05$ as compared to the old OVX rats treated with 17β-estradiol (17β-E_2). Each group comprised a minimum of eight rats. Cholecalciferol was given at 1.0, 2.5 or 5.0 mg/kg/day subcutaneously (s.c.), once daily, for 14 days. The administered dose of 17β-estradiol was 0.5 µg/rat s.c., once daily, for 14 days.*

(**Figure** 7, $p > 0.05$). Cholecalciferol at all doses in combination with 17β-E_2 significantly increased 25-OH-VD_3 levels when these old OVX rats were compared with the old OVX/solvent and OVX/17β-E_2 groups (**Figure 8**, $p > 0.05$).

The two-way ANOVA failed to show any significant differences in the calcium levels in the blood serum between hormone conditions, drug treatments, and an interaction between hormone condition and treatments in the middle-aged and old OVX rats with long-term estrogen deficiency (**Figures 9** and **10**, $p > 0.05$). The *post-hoc* test did not find any differences among the experimental groups for the calcium levels ($p > 0.05$).

4. Discussion

In the present study, the effects of chronic cholecalciferol treatment at different doses (1.0, 2.5 and 5.0 mg/kg, s.c.) for 14 days on anxiety-like behavior in the middle-aged and old female rats with long-term estrogen deficiency and 17β-E_2 supplementation in a low dose were examined. Endogenous estrogens were removed by ovariectomy and only after 12 weeks post-ovariectomy period, these rats were used in all experiments. The results of behavioral testing for the anxiety-related effects of cholecalciferol were compared in both old OVX rats and OVX female rats treated with 17β-E_2. Simultaneously, the effects of cholecalciferol at similar doses on anxiety-like behavior were tested in middle-aged and old intact female rats. For this purpose, the elevated plus maze (EPM) and light-dark test (LDT) were made use in the present study. It was also investigated whether the effects of cholecalciferol at different doses were specific in the EPM and LDT, measuring its effects on the behavioral activity in the OFT of the middle-aged and old intact and OVX rats after long-term absence of estrogen.

Cholecalciferol at all investigated doses did not produce any changes of anxiety-like behavior of the middle-aged and old intact female rats in the EPM and LDT. Analyzing the results from biochemical assay, an increase of 25-OH-VD_3 concentrations and absence of any modifications of estradiol levels in the serum blood of the middle-aged and old intact rats given with different doses of cholecalciferol were found.

These results suggest that cholecalciferol induced the increasing of 25-OH-VD_3 levels in the blood serum of the middle-aged and old intact-ovary rats are not associated with absence of anxiety-like profile alterations in the behavioral tests.

Furthermore, the ovary-intact female rats of different age are also needed to evaluate the behavioral effects of cholecalciferol administered at several doses in the EPM and LDT paradigms.

The results showed that in the middle-aged and old OVX rats following 12 weeks of post-ovariectomy period, there were marked anxiety-like behavior as assessed by EPM and LDT. Although 17β-E_2 supplementation resulted in significant anxiolytic-like effect of the middle-aged and old OVX rats with long-term absence of estrogen, the 17β-E_2 administration was not able to completely diminish anxiety-like behavior to the level of the middle-aged and old control intact animals. According to these results, we conclude that middle-aged and old OVX rats following 12 weeks of postovariectomy period display significant anxiety-related behavior, while 17β-E_2 administration to the middle-aged and old OVX rats attenuates the estrogen deficiency-induced anxiety-like behavior to some extent. In fact, these experiments showed that the effects of 17β-E_2 supplementation on anxiety-like behavior did not associated with absence of its effects on 25-OH-VD_3 levels in the old OVX rats. The long-term effect of ovariectomy on anxiety-like behavior in female rats of different age that were submitted in a standard behavioral tests [24, 35, 36].

Cholecalciferol at dose of 5.0 mg/kg/day per se had a significant anxiolytic-like effect in the middle-aged OVX rats following long-term ovariectomy. On the contrary, cholecalciferol at doses of 1.0 and 2.5 mg/kg/day failed to induce any modifications of anxiety-like behavior in the middle-aged OVX rats with long-term absence of estrogen. Combined application of vitamin D_3 (1.0 mg/kg/day) with 17β-E_2-induced synergic anxiolytic-like action of both preparations in the anxiety-related tasks of the old OVX rats. Cholecalciferol at all doses per se had a significant anxiolytic-like effect in the old OVX rats following long-term ovariectomy. Unexpectedly, that the old OVX rats with 12 weeks post-ovariectomy period administered with cholecalciferol at doses 2.5 and 5.0 mg/kg/day in combination with 17β-E_2 showed similar anxiety-like profile like the OVX rats given with 17β-E_2. Thus, we did not observe any synergic anxiolytic-like effects of cholecalciferol at doses of 2.5 or 5.0 mg/kg in the old OVX rats given with 17β-$E_{2.}$ It might suppose that there are some concurrent relation between 17β-E_2 and cholecalciferol at doses of 2.5 and 5.0 mg/kg/day. In fact, application of 17β-E_2 interfere with anxiolytic-like action of cholecalciferol at doses of 2.5 or 5.0 mg/kg in the old OVX rats after long-term ovariectomy. Simultaneously, cholecalciferol treatment in all tested doses similarly increased grooming, did not change locomotor activity and rearing of the middle-aged and old OVX rats after long-term ovariectomy. Thus, the present results suggest that anxiolytic-like effects of vitamin D_3 are specific in the OVX rats of different age given with solvent or 17β-E_2, since data of the OFT were not able to demonstrate any of alterations in motor or rearing activities in these rats. Combined application of vitamin D_3 (5.0 mg/kg/day) with low dose of 17β-E_2 synergistically decreased anxiety-like profile of the middle-aged OVX rats. However, vitamin D_3 at all doses significantly decreased grooming behavior in the middle-aged OVX rats treated with 17β-E_2.

Unexpectedly, the middle-aged OVX rats treated with vitamin D_3 at doses 1.0 and 2.5 mg/kg/day in combination with 17β-E_2 have practically identical parameters of anxiety-like state as the OVX rats received only 17β-E_2. Thus, this study failed to show any anxiolytic-like effects of vitamin D_3 at doses of 1.0 or 2.5 mg/kg in the middle-aged OVX treated with 17β-$E_{2.}$ The present results might speculate that there exists some concurrent relation between 17β-E_2 and vitamin D_3 at doses of 2.5 and 5.0 mg/kg/day for the brain structures implicated in the mechanisms of anxiety. In fact, additional application of 17β-E_2 to the middle-aged OVX rats interfered with anxiolytic-like action of vitamin D_3 when it administered at doses 1.0 or

2.5 mg/kg. Thus, the effects of vitamin D_3 administration in a combination with low dose of 17β-E_2 on anxiety-like state of the middle-aged OVX rats are determined by dose of treatment.

ELISA assays demonstrated that administration of cholecalciferol only at doses of 2.5 and 5.0 mg/kg resulted in elevated 25-OH-VD_3 levels in the blood serum of the middle-aged OVX rats with long-term absence of estrogen. Moreover, application of cholecalciferol at these doses with low dose of 17β-E_2 induced more profound increase of 25-OH-VD_3 levels in the blood serum of the middle-aged OVX rats. These data suggest that the different effects of cholecalciferol application in the middle-aged OVX rats with long-term absence of estrogen on anxiety-like behavior in the EPM and LDT did not associated with its effects on 25-OH-VD_3 levels.

Biochemical analysis showed that administration of cholecalciferol at all doses alone or in a combination with 17β-E_2 resulted in elevated 25-OH-VD_3 levels in the blood serum of the old OVX rats with long-term absence of estrogen. Cholecalciferol administered alone at all doses similarly increased estradiol levels in the blood serum of the old OVX rats after long-term ovariectomy. On the other hand, only application of cholecalciferol at a dose of 1.0 mg/kg with low dose of 17β-E_2 induced more profound increase of estradiol levels in the blood serum of the old OVX rats. Moreover, cholecalciferol in several doses failed to induce any changes in calcium concentrations in the blood serum of the old OVX rats given with solvent or 17β-E_2.

These data suggest that the different effects of cholecalciferol application per se in the middle-aged and old OVX rats with long-term absence of estrogen on anxiety-like behavior in the EPM and LDT did not associated with its effects on estradiol, 25-OH-VD_3 and calcium levels.

However, we can speculate that behavioral effects of cholecalciferol treatment with low dose of 17β-E_2 in the EPM and LDT tests might connected with fluctuations of estradiol levels in the blood serum of the middle-aged and old OVX rats. It is possible that specific sites of action involved in the anxiolytic-like effects of cholecalciferol that also modulated by estrogens are affected by the endocrine milieu that prevails at different period for the middle-aged and old female rats. Moreover, after a long-time absence of ovarian fluctuations an adaptive process may contribute to a better response for cholecalciferol administration at a dose of 1.0 mg/kg in the old female rats and for cholecalciferol at a dose of 5.0 mg/kg in the middle-aged female rats.

The role of ovarian hormones in anxiety and stress sensitivity is of great interest for women transitioning through menopause [37, 38]. Mood disorders during menopause could be partly due to loss of estrogen with menopause because estrogen is known to have neuroprotective effects on brain [39, 40]. Menopausal hormonal therapy (MHT) may improve the symptoms of affective-related state in people or decrease the risk of developing mood disturbances in older women, but this is unclear because in some studies MHT does not stop the development of anxiety-like symptoms in elderly postmenopausal women [41]. The exact role of estrogen still needs to be defined.

Menopause are also at higher risk of developing VD deficiency due to decreased dietary intake, less sun exposure, restricted outdoor activity and a decreased capacity to produce enough calcitriol as a result of an age related decline in hydroxylation by kidneys [42–44].

Vitamin D_3 is a hormone precursor which is transformed into 1,25-dihydroxyvitamin D (1,25-$(OH)_2D_3$) in the liver and kidney [45]. Through decades, this active form of VD has been involved in the brain development and functions of the central nervous system (CNS) [46, 47]. Hormonal form 1,25(OH)2D_3 enters the brain via the blood-brain barrier to act directly on cells containing its nuclear receptor, the VDR [48, 49].

The presence of VD receptors (VDR) outside the skeletal system, enterocytes and renal tubular cells was confirmed in many cell types including immune cells, neurons, pancreatic cells, myocytes, cardiomyocytes, endothelium cells, which stress pleiotropic activity of VD [50, 51]. The active form of vitamin D is transferred to astrocytes where it can bind to VDR and initiate gene transcription or be inactivated when in excess [48, 49]. Alternatively, $1,25(OH)_2D_3$ can induce autocrine or paracrine rapid non-genomic actions since all brain cell types express the other membrane receptor of VD [48]. It is possible that the behavioral effects of cholecalciferol are mediated by multiple target regions, including brain centers that are involved in the mechanisms of anxiety-like behavior. Regardless, it cannot presently exclude possible indirect effects of vitamin D_3 on different neurotransmitter circuits. Further research is required to investigate the underlying mechanisms in its anxiolytic-like effects.

It is well-established both systemic effects of VD on calcium metabolism and neuroprotective effects of VD on the brain [52]. Low VD levels has been implicated in the pathophysiological mechanisms of cardiovascular diseases, depression, anxiety, cognitive disorders, obesity, metabolic syndrome, type 2 diabetes, various types of cancer and immune disorders [10, 52]. According to Gaugris and co-workers [53], the prevalence of low VD levels appears to be high in postmenopausal women. Additionally, the decline of estrogens after menopause decreases the activity of 1α-OHase, which results in lower synthesis of the active VD form [53, 54]. Application of vitamin D_3 in specific periods of women's life, seems to be of great importance, because it may reduce the risk of affective-related diseases during menopausal period [55–57]. However, the current data for vitamin D_3 application studies are very incomplete and need of more intensive investigations. The main point of question is to examine how the interaction between vitamin D_3 and estradiol might alter at specific periods of women's life, and the impacts of such alterations elsewhere in the postmenopausal woman. VDR have been identified throughout the female reproductive tract [58, 59]. Some studies have demonstrated a direct modulation by VD of estradiol, estrone and progesterone production in human ovarian cells [60–62]. It could be supposed that estrogens and VD share similar targets of the brain to induce their anxiolytic-like effects. However, the behavioral manifestations of VD at various doses are completely different in the middle-aged and old OVX females. It is likely that VD acts through a various mechanisms that are sensitive in female rats of different age with long-term absence of ovarian hormones. Moreover, it is completely needed to understand the precise mechanisms of how VD treatment alone or in a combination with 17β-estradiol supplementation may affect women's anxiety-related state.

These points illustrate how the current state of VD treatment research is incomplete and in need of more intensive research. Working toward uncovering how the interaction between VD and estradiol changes after menopause, and the implications of these changes elsewhere in the postmenopausal woman, is necessary for providing the most complete understanding of how VD treatment alone or in a combination with 17β-estradiol supplementation may affect women's affective-related state.

VD as changes in VDR impact on various brain neurotransmitters, and thus suggest a potential role of vitamin D in causing and redressing mood disorders [63]. It could be supposed, even though estrogens and cholecalciferol share similar targets on monoaminergic or another neurotransmitter system to induce their anxiolytic-like effects, the behavioral manifestations of cholecalciferol are completely different in the old OVX females. It is likely that cholecalciferol acts through a different mechanisms of action that is sensitive to the age of female rats with long-term absence of ovarian hormones.

In conclusion, the results of this study can be summarized as follows: all tested doses of cholecalciferol given alone are produced anxiolytic-like effects in the

EPM and LDT in the old OVX female rats; the one specific dose of cholecalciferol (1.0 mg/kg/day) in the old OVX rats and another one dose of cholecalciferol (5.0 mg/kg/day) in the middle-aged OVX rats are able to induce synergic anxiolytic-like effect in the EPM and LDT; effects of cholecalciferol on anxiety-like behavior in the middle-aged and old OVX rats after long-term absence of estrogen are dependent from absence or presence of additional hormonal treatment as 17β-E_2. Further investigations is to be addressed in relation to such issues: whether different effects of cholecalciferol on anxiety-like behavior are dependent from different age of rats, or whether different doses of cholecalciferol on anxiety-like behavior in OVX rats with different age rats might lead to negative versus positive effects. Further research, with properly designed experimental studies, is needed to test this hypothesis. In addition, further research is needed to elucidate the biochemical mechanism/(s) of cholecalciferol effects on the anxiety-like behavior and their physiological relevance for development of mood impairment at estrogen deficiency at aging. Furthermore, the mechanism by which cholecalciferol produces anxiolytic-like effect in the middle-aged and old OVX rats and the implications of this in brain function need to be investigated in future research. Moreover, further studies are needed to evaluate the association of VD with estrogen-related pathways and to conduct detail experiments together with biochemical studies of these subjects to verify the significance of this study.

5. Conclusions

The present data of the preclinical study indicates that chronic cholecalciferol at a dose of 5.0 or 1.0 mg/kg, s.c. treatment decreased anxiety-related behavior after impairment induced by long-term ovariectomy in the middle-aged and old female rats with long-term absence of estrogen, respectively. The data also indicate that the combination of cholecalciferol at a dose of 5.0 mg/kg and 17β-E_2 is more effective than 17β-E_2 alone in the middle-aged OVX rats inducing more synergic anxiolytic-like effects in the EPM and LDT. Moreover, a combination of cholecalciferol at a dose of 1.0 mg/kg s.c. and 17β-E_2 is more effective than 17β-E_2 alone in the old OVX rats inducing a more synergic anxiolytic-like effects in the EPM and LDT. Furthermore, this is the first study to show a beneficial effect of chronic cholecalciferol at dose of 1.0 mg/kg s.c. administration on anxiety-related states induced by long-term ovariectomy in the old female rats. Importantly, these results suggest that 17β-E_2 administration interfered with anxiolytic-like action of cholecalciferol administered alone at doses of 2.5 or 5.0 mg/kg to the old OVX rats with long-term absence of estrogens. This work promotes more effective creating of the novel therapeutic targets and strategies for anxiety-like treatment in the middle-aged and old subjects with long-term estrogen deficiency.

Acknowledgements

The reported study was funded by Russian Science Foundation (RSF) accordingly to the research project No. 16-15-10053.

Conflict of interest

The author declares no conflict of interest.

Author contributions

Author Julia Fedotova designed the study, performed the experiments, analyzed the data and wrote the protocol and manuscript.

Author details

Julia Fedotova[1,2]

1 Laboratory of Neuroendocrinology, I.P. Pavlov Institute of Physiology of the Russian Academy of Sciences, St. Petersburg, Russia

2 Faculty of Food Biotechnologies and Engineering, ITMO University, St. Petersburg, Russia

*Address all correspondence to: julia.fedotova@mail.ru

References

[1] Terauchi M, Hiramitsu S, Akiyoshi M, Owa Y, Kato K, Obayashi S, et al. Associations between anxiety, depression and insomnia in peri- and postmenopausal women. Maturitas. 2012;**72**:61-65. DOI: 10.1016/j.maturitas.2012.01.014

[2] Castanho TC, Moreira PS, Portugal-Nunes C, Novais A, Costa PS, Palha JA, et al. The role of sex and sex-related hormones in cognition, mood and well-being in older men and women. Biological Psychology. 2014;**103**:158-166. DOI: 10.1016/j.biopsycho.2014.08.015

[3] Soares CN, Maki PM. Menopausal transition, mood, and cognition: An integrated view to close the gaps. Menopause. 2010;**17**:812-814. DOI: 10.1097/gme.0b013e3181de0943

[4] Maki PM, Freeman EW, Greendale GA, Henderson VW, Newhouse PA, Schmidt PJ, et al. Summary of the National Institute on aging-sponsored conference on depressive symptoms and cognitive complaints in the menopausal transition. Menopause. 2010;**17**:815-822. DOI: 10.1097/gme.0b013e3181d763d2

[5] Arevalo M-A, Azcoitia I, Garcia-Segura LM. The neuroprotective actions of oestradiol and oestrogen receptors. Nature Reviews Neuroscience. 2015;**16**:17-29. DOI: 10.1038/nrn3856

[6] Soares CN. Mood disorders in midlife women: Understanding the critical window and its clinical implications. Menopause. 2014;**21**:198-206. DOI: 10.1097/GME.0000000000000193

[7] Walf AA, Frye CA. ER(beta)-selective estrogen receptor modulators produce antianxiety behavior when administered systemically to ovariectomized rats. Neuropsychopharmacology. 2005;**30**:1598-1609

[8] Soares CN. Depression in peri- and postmenopausal women: Prevalence, pathophysiology and pharmacological management. Drugs & Aging. 2013;**30**:677-685. DOI: 10.1007/s40266-013-0100-1

[9] Soares CN, Almeida OP, Joffe H, Cohen LS. Efficacy of estradiol for the treatment of depressive disorders in perimenopausal women: A double-blind, randomized, placebo-controlled trial. Archives of General Psychiatry. 2001;**58**:529-534

[10] Pae CU, Mandelli L, Kim TS, Han C, Masand PS, Marks DM, et al. Effectiveness of antidepressant treatments in pre-menopausal versus post-menopausal women: A pilot study on differential effects of sex hormones on antidepressant effects. Biomedicine & Pharmacotherapy. 2009;**63**:228-235. DOI: 10.1016/j.biopha.2008.03.010

[11] Vedder LC, Bredemann TM, McMahon LL. Estradiol replacement extends the window of opportunity for hippocampal function. Neurobiology of Aging. 2014;**35**:2183-2192. DOI: 10.1016/j.neurobiolaging.2014.04.004

[12] Genaro PS, Pereira GAP, Pinheiro MM, Szejnfeld VL, Martini LA. Relationship between nutrient intake and vitamin D status in osteoporotic women. International Journal for Vitamin and Nutrition Research. 2007;**77**:376-381. DOI: 10.1024/0300-9831.77.6.376

[13] Scheid V, Ward T, Ch WS, Watanabe K, Liao X. The treatment of menopausal symptom, s by traditional east Asian medicines: Review and perspectives. Maturitas. 2010;**66**:111-130. DOI: 10.1016/j.maturitas.2009.11.020

[14] Studd J, Nappi RE. Reproductive depression. Gynecological

Endocrinology. 2012;**28**(Suppl. S1):42-45. DOI: 10.3109/09513590.2012.651932

[15] Peng W, Sibbritt DW, Hickman L, Adams J. Association between use of self-prescribed complementary and alternative medicine and menopause-related symptoms: A cross-sectional study. Complementary Therapies in Medicine. 2016;**23**:666-673

[16] Holick MF, Frommer JE, McNeill SC, Richtand NM, Henley JW, Potts JT Jr. Photometabolism of 7-dehydrocholesteroltoprevitamin D3 in skin. Biochemical and Biophysical Research Communications. 1977;**76**:107-114

[17] Groves NJ, McGrath JJ, Burne TH. Vitamin D as a neurosteroid affecting the developing and adult brain. Annual Review of Nutrition. 2014;**34**:117-141. DOI: 10.1146/annurev-nutr-071813-105557

[18] Christakos S, Raval-Pandya M, Wernyj RP, Yang W. Genomic mechanisms involved in the pleiotropic actions of 1,25-dihydroxyvitamin D3. The Biochemical Journal. 1996;**316**:361-371

[19] Mizwicki MT, Norman AW. The vitamin D sterol-vitamin D receptor ensemble model offers unique insights into both genomic and rapid-response signaling. Science Signaling. 2009;**2**:re4. DOI: 10.1126/scisignal.275re4

[20] Cui X, McGrath JJ, Burne TH, Mackay-Sim A, Eyles DW. Maternal vitamin depletion alters neurogenesis in the developing rat brain. International Journal of Developmental Neuroscience. 2007;**25**:227-232

[21] De Luca HF. History of the discovery of vitamin D and its active metabolites. Bone Key Reports. 2014;**3**:479. DOI: 10.1016/j.ijdevneu.2007.03.006

[22] Eyles DW, Smith S, Kinobe R, Hewison M, McGrath JJ. Distribution of the vitamin D receptor and 1a-hydroxylase in human brain. Journal of Chemical Neuroanatomy. 2005;**29**:21-30

[23] De Chaves G, Moretti M, Castro AA, Dagostin W, da Silva GG, Boeck CR, et al. Minireview: Vitamin D: Is there a role in extraskeletal health? Endocrinology. 2011;**152**:2930-2936

[24] Estrada-Camarena E, Lopez-Rubalcava C, Azucena Hernandez-Aragon A, Silvia Mejıa-Mauries S, Picazo O. Long-term ovariectomy modulates the antidepressant-like action of estrogens, but not of antidepressants. Journal of Psychopharmacology. 2017;**25**:1365-1377

[25] Bosee R, Di Paolo T. Dopamine and GABAA receptor imbalance after ovariectomy in rats: Model of menopause. Journal of Psychiatry & Neuroscience. 1995;**20**:364-371

[26] Estrada-Camarena E, Fernandez-Guasti A, Lopez-Rubalcava C. Interaction between estrogens and antidepressants in the forced swimming test in rats. Psychopharmacology. 2004;**173**:139-145. DOI: 10.1007/s00213-003-1707-4

[27] Idrus NM, Happer JP, Thomas JD. Cholecalciferol attenuates perseverative behavior associated with developmental alcohol exposure in rats in a dose-dependent manner. The Journal of Steroid Biochemistry and Molecular Biology. 2013;**136**:146-149

[28] Estrada-Camarena E, Fernandez-Guasti A, Lopez-Rubalcava C. Antidepressant-like effect of different estrogenic compounds in the forced swimming test.

Neuropsychopharmacology. 2003;**28**:830-838. DOI: 10.1038/sj.npp.1300097

[29] Pellow S, File SE. Anxiolytic and anxiogenic drug effects on exploratory activity in an elevated plus-maze: A novel test of anxiety in the rat. Pharmacology, Biochemistry, and Behavior. 1986;**24**(3):525-529. DOI: 10.1016/ 0091-3057(86)90552-6

[30] Pellow S, Chopin P, File SE, Briley M. Validation of open: Closed arm entries in an elevated plus-maze as a measure of anxiety in the rat. Journal of Neuroscience Methods. 1985;**14**:149-167

[31] Menzaghi F, Howard RL, Heinrichs SC, Vale W, Rivier J, Koob GF. Characterization of a novel and potent corticotropin-releasing factor antagonist in rats. The Journal of Pharmacology and Experimental Therapeutics. 1994;**269**:564-572

[32] Edinger KL, Frye CA. Sexual experience of male rats influences anxiety-like behavior and androgen levels. Physiology & Behavior. 2007;**92**:443-453

[33] Pan H-Z, Chen H-H. Hyperalgesia, low-anxiety, and impairment of avoidance learning in neonatal caffeine-treated rats. Psychopharmacology. 2006;**191**:119-125

[34] Fedotova J, Soultanov V, Nikitina T, Roschin V, Ordayn N. Ropren® is a polyprenol preparation from coniferous plants that ameliorates cognitive deficiency in a rat model of beta-amyloid peptide-(25-35)-induced amnesia. Phytomedicine. 2012;**19**:451-456. DOI: 10.1016/j.phymed.2011.09.073

[35] Nelly M, Rivera V, Tenorio AG, Fernández-Guasti A, Estrada-Camarena E. The post-ovariectomy interval affects the antidepressant-like action of citalopram combined with ethynyl-estradiol in the forced swim test in middle aged rats. Pharmaceuticals. 2016;**9**:21

[36] Okada M, Hayashi N, Kometani M, Nakao K, Inukai T. Influences of ovariectomy and continuous replacement of 17β-estradiol on the tail skin temperature and behavior in the forced swimming test in rats. Japanese Journal of Pharmacology. 1997:93-96

[37] Burger H. The menopausal transition-endocrinology. The Journal of Sexual Medicine. 2008;**5**:2266-2273. DOI: 10.1111/j.1743-6109.2008.00921.x

[38] MacLennan AH, Taylor AW, Wilson DH. Hormone therapy use after the women's health initiative. Climacteric. 2004;**7**:138-142

[39] Przybelski R, Binkley N. Is vitamin D important for preserving cognition? A positive correlation of serum 25-hydroxyvitamin D concentration with cognitive function. Archives of Biochemistry and Biophysics. 2007;**460**:202-220

[40] Wilkins C, Sheline Y, Roe C, Birge S, Morris J. Vitamin D deficiency is associated with low mood and worse cognitive performance in older adults. The American Journal of Geriatric Psychiatry. 2006;**14**:1032-1040

[41] Rossouw JE, Anderson GL, Prentice RL, LaCroix AZ, Kooperberg C, Stefanick ML, et al. Risks and benefits of estrogen plus progestin in healthy postmenopausal women: Principal results from the women's health initiative randomized controlled trial. Journal of the American Medical Association. 2002;**288**:321-333

[42] Cheema C, Grant BF, Marcus R. Effects of estrogen on circulating "free" and total 1,25-dihydroxyvitamin

D and on the parathyroid-vitamin D axis in postmenopausal women. The Journal of Clinical Investigation. 1989;**83**:537-542. DOI: 10.1172/JCI113915

[43] Schnatz PF, Marakovits KA, O'Sullivan DM, Ethun K, Clarkson TB, Appt SE. Response to an adequate dietary intake of vitamin D3 modulates the effect of estrogen therapy on bone density. Journal of Women's Health (2002). 2012;**21**:858-864. DOI: 10.1089/jwh.2011.3244

[44] Robbins JA, Aragaki A, Crandall CJ, Manson JE, Carbone L, Jackson R, et al. Women's health initiative clinical trials: Interaction of calcium and vitamin D with hormone therapy. Menopause. 2014;**21**:116-123. DOI: 10.1097/GME.0b013e3182963901

[45] Stewart A, Wong K, Cachat J, Elegante M, Gilder T, Mohnot S, et al. Neurosteroid vitamin D system as a nontraditional drug target in neuropsychopharmacology. Behavioural Pharmacology. 2010;**21**:420-426. DOI: 10.1097/FBP.0b013e32833c850f

[46] Holick MF. High prevalence of vitamin D inadequacy and implications for health. Mayo Clinic Proceedings. 2006;**81**:353-373

[47] Penckofer S, Kouba J, Byrn M, Estwing Ferrans C. Vitamin D and depression: Where is all the sunshine? Issues in Mental Health Nursing. 2010;**31**:385-393. DOI: 10.3109/01612840903437657

[48] Eyles DW, Burne TH, McGrath JJ. Vitamin D, effects on brain development, adult brain function and the links between low levels of vitamin D and neuropsychiatric disease. Frontiers in Neuroendocrinology. 2013;**34**:47-64. DOI: 10.1016/j.yfrne.2012.07.001

[49] Eyles DW, Liu PY, Josh P, Cui X. Intracellular distribution of the vitamin D receptor in the brain: Comparison with classic target tissues and redistribution with development. Neuroscience. 2014;**268**:1-9. DOI: 10.1016/j.neuroscince.2014.02.042

[50] Holick MF. Vitamin D deficiency. New England Journal of Medicine. 2007;**357**:266-281

[51] Drevets WC, Price JL, Furey ML. Brain structural and functional abnormalities in mood disorders: Implications for neurocircuitry models of depression. Brain Structure and Function. 2008;**213**:93-118. DOI: 10.1007/s00429-008-0189x

[52] Fernandes de Abreu DA, Eyles D, Féron F. Vitamin D, a neuro-immunomodulator: Implications for neurodegenerative and autoimmune diseases. Psychoneuroendocrinology. 2009;**34**:S265-S277. DOI: 10.1016/j.psyneuen.2009.05.023

[53] Gaugris S, Heaney RP, Boonen S, Kurth H, Bentkover JD, Sen SS. Vitamin D inadequacy among post-menopausal women: A systemic review. QJM: An International Journal of Medicine. 2005;**98**:667-676

[54] Bikle DD. Vitamin D metabolism, mechanism of action and clinical applications. Chemistry and Biology. 2014;**21**:319-329. DOI: 10.1016/j.chembiol.2013.12.016

[55] Garcion E, Wion-Barbot N, Montero-Menei C, Berget F, Wion D. New clues about vitamin D functions in the nervous system. Trends in Endocrinology and Metabolism. 2005;**13**:100-105

[56] Kalueff AV, Eremin K, Tuohimaa P. Mechanisms of neuroprotective action of vitamin D3. The Biochemist. 2004;**69**:738-741

[57] Eyles D, Brown J, Mackay-Sim A, McGrath J, Feron F. Vitamin D3 and brain development. Neuroscience. 2003;**118**:641-653

[58] Halloran BP, De Luca HF. Effect of vitamin D deficiency on fertility and reproductive capacity in the female rat. The Journal of Nutrition. 1980;**110**:1573-1580

[59] Kwiecinksi GG, Petrie GI, De Luca HF. 1,25-Dihydroxyvitamin D3 restores fertility of vitamin D-deficient female rats. The American Journal of Physiology. 1989;**256**:E483-E487. DOI: 10.1152/ajpendo.1989.256.4.E483

[60] Kinuta K, Tanaka H, Moriwake T, Aya K, Sato S, Seino Y. Vitamin D is an important factor in estrogen biosynthesis of both female and male gonads. Endocrinology. 2004;**141**:1317-1324. DOI: 10.1210/endo.141.4.7403

[61] Luk J, Torrealday S, Neal Perry G, Pal L. Relevance of vitamin D in reproduction. Human Reproduction (Oxford, England). 2012;**27**:3015-3027. DOI: 10.1093/humrep/des248

[62] Ozkan S, Jindal S, Greenseid K, Shu J, Zeitlian G, Hickmon C, et al. Replete vitamin D stores predict reproductive success following in vitro fertilization. Fertility and Sterility. 2010;**94**:1314-1319. DOI: 10.1016/j.fertnstert.2009.05.019

[63] Kiraly SJ, Kiraly MA, Hawe RD, Makhani V. Vitamin D as neuroactive substance: Review. The Scientific World Journal. 2006;**6**:125-139. DOI: 10.1100/tsw.2006.25

Section 3

New Frontiers for Vitamin D Research

Chapter 5

Vitamin D in Space

Amir Khoshvaghti

Neil Armstrong stepped on the Moon and said:

"That's one small step for [a] man, one giant leap for mankind." [1]

However, it was better to say:

"That's one giant step for [a] man, one small leap for mankind."

Abstract

Mankind has explored space since many years ago for distinct purposes. The space environment has its special features including microgravity (weightlessness), radiation, vacuum, and extreme temperature. It is fascinating to hear space traveling or even living on another planet, but it may cause dramatic changes in the human body. The skeleton has the primary role for the human body on the Earth and also in space. Osteoporosis is the principal feature in spaceflights occurring sooner or later. It will pose health risks. The aging of the population is the cause for osteoporosis to be more prevalent in the future. It seems that aging will happen sooner in space for human beings that are accustomed to living on the Earth. A complex of all reported changes that would occur in space, the picture of a closely resembling aged man emerges. Vitamin D can be synthesized and acts as a hormone. Its receptors are evident in more than 30 human tissues. Vitamin D has positive effects on the skeleton and other systems of the body in many regards. Multiple actions have been claimed for vitamin D (real or false), many aspects of which are not fully understood. The issue applies more about the vitamin and space, which has been tried to describe in this chapter.

Keywords: vitamin D, microgravity, spaceflight, bone loss, health risks

1. Introduction

The humankind had long been thinking and desiring to separate from the Earth. He lived on the Earth but considered it as a prison. Separating from the ground and climbing up for reaching the blue and bright sky were a kind of evolution and progression for a man's dreams. One of the reasons for this thought was the natural curiosity of human being, along with other reasons including scientific progress, financial outcome, and so on.

2. The desire to fly

The humankind in this field even turned to the imagination. He always dreamed about the angels, carrying two flights on the back of them. He made stories and myths like Icarus about flying and releasing to the sky.

Problems gradually showed up themselves. The first problem was the separation from the Earth. Every time that he jumped up, he fell on the ground again, even from towers. Each time he tried to get higher, the gravity did not allow him to move away from the Earth, and he was returned. He then realized that he had to use a special force to win over and dominate the gravity. Little by little he took more serious actions, designed to make practical models of flight. The Chinese invented a rocket, which had the necessary power for overcoming the gravity. Leonardo da Vinci was a brilliant scientist, designer, and inventor of the field in the fifteenth century.

A Journey through Space was written by William Leitch in 1861. He was a Scottish astronomer and mathematician. He announced the theory of traveling to space with rockets [2].

Russian scientist Konstantin Eduardovich Tsiolkovsky (1857–1935) wrote about going to space using special devices in 1903 (the same year as the Wright brothers' flight) [2].

At last, the man could be separated from the Earth to fly in the sky. The German Otto Lilienthal (1848–1896) studied flying scientifically and had 200 gliding flights in the second half of the nineteenth century [3]. The Wright brothers (Orville, 1871–1948 and Wilbur, 1867–1912) as American aviation pioneers made the first piloted flights in 1903 [4]. Now it was the time to provide suitable facilities to fly longer or to reach higher.

The past Soviet Union (current Russia) sent the first human (Yuri Gagarin, 1934–1968) to the orbit on April 12, 1961. They had sent animals to space before this exciting experiment. The first animal was a Russian dog called Laika 1 month after the launch of Sputnik on November 3, 1957. Russians sent six dogs to the orbit from 1960 to 1961. A chimpanzee called Enos was the first animal sent to space by the USA on November 29, 1961. John Glenn was the first American astronaut in space on February 20, 1962 [5].

Since the flight of Yuri Gagarin, more than 500 people have flown to the orbit of the Earth. Even a few have advanced beyond the orbit, the Moon.

After starting to fly, humans realized that there would be more problems at higher altitudes. One problem became evident when he went a bit higher. The atmosphere became thinner and thinner, so he was forced to struggle for breathing. Then he had to make the flight device so sealed that it could regulate the air inside to have no respiratory problems or other problems due to lack of oxygen and hypoxia. He designed and made special balloons for flying at high altitudes (rediscovery of hot air balloon by the Montgolfier brothers) [6].

At the same time, scientists noticed that everything could not be experienced in the sky (or space) and during flight. Unfortunately, some experiences led to fall, collapse, and death. The experiments were commenced on the Earth by simulating different models (from cells to animals and finally up to human beings).

At that time, the new problems appeared one after another. The man noticed that by going farther from the surface of the Earth, there would be less gravity. Somewhat farther, he almost did not feel any gravity. Weightlessness had occurred.

Other problems were included as exposure to radiation and its resulting discomforts. So, on the one hand, the man had to think about the ways continuing to fly and perform missions successfully and, on the other hand, not to confront any physical problem himself, doing his best with maximum efficiency and power.

The efficient hypothesis was that the astronaut should be able to maintain his abilities in long-term flights up to the end and to act effectively meanwhile. In recent decades, 6-month missions have been carried out regularly in Russia's Salyut, the Mir station, and the International Space Station (ISS), respectively [7]. In addition to such missions, various studies have been done on human body changes in space travels and missions, in order to grasp the practical concept of adaptive conditions, although this understanding is still incomplete.

The space environment and space travel have particular challenges and dangers within itself. Regular and planned efforts have been made to understand these changes and alterations in order to allow protection and countermeasures development and implementation against harmful effects. The field of optimization of human performance in every respect is provided. In this context, the following should be considered:

1. Changes in acceleration at two steps:

    a. When the spacecraft rises from the ground

    b. At landing on the surface of other planets or upon returning to the Earth (from full extent until its complete absence during weightlessness)

2. Different atmosphere compositions (containing variable pressures and different amounts of oxygen)

3. Other features such as radiation (including ionizing radiation)

It is nearly two decades since the establishment of the first component of ISS in 1998. It is believed to operate until 2028. Several long-term missions have been planned to accommodate astronauts. The main goal is to get the ultimate readiness for exploratory missions beyond the Earth's orbit [8].

The American and Russian space programs were ambitious and, at the same time, scientific and full of research. Their main goal was the long-term deployment of systems and platforms in the Earth's orbit to make orbit stations. They intended to send astronauts to these platforms and stations for long-duration missions. Astronauts had to stay and work at space stations for a long time so that reliable research could be done in this area. Scientists achieved much more about physiological and pathological changes in the space. The complete adaptation of the human body to weightlessness would take several weeks, which could not be provided by previous spaceships [9]. Also, many experiments and researches were possible to be done in this case.

Sometimes later, the scientific thoughts looked farther, beyond the Earth's orbit. Terrestrial missions refer to those space missions that are far from the orbit, where there is no longer any protective effect of the atmosphere and the electromagnetic fields, and they will not affect the spacecraft longer. The American Apollo programs are among these series of missions that consisted of nine missions and flights to the Moon. In six missions, two astronauts landed on the surface of the Moon. Each trip lasted between 8 and 12 days, which spent 3 days traveling from the Earth to the Moon. An essential point about Apollo's trip was the ability of humans to perform physical activities in an environment with gravity much lower than on the Earth's surface, equivalent to one-sixth. It is worthy to mention that gravitational force on the surface of Mars is twice that of the Moon and a third of the Earth [10].

With the advent of space exploration by Vostok 1 and Yuri Alekseyevich Gagarin's journey in 1961, the man has been exposed to space-related problems including microgravity [2]. The cells, tissues, systems, and the human body as a whole are generally exposed to different physiology other than the physiology that dominates our planet and undergoes alterations. Of course, the musculoskeletal system is among those systems that have the most changes and problems, especially in the face of microgravity, such as bone mineral density decrease and muscle atrophy [11].

Microgravity and cosmic radiation are two important dilemmas that exist in the space environment affecting human health and endangering it. They significantly affect the future of long-term space travels and would limit them [12].

In a long space trip, including a trip to Mars, astronauts are exposed to large amounts of radiation which is much higher than the magnitude permitted by the NASA's system of protection [13].

3. History

In the twentieth century, spaceflight became real. The modern space exploration era begins in 1961 with Vostok 1 and Yuri Gagarin's journey because human travel to space is the most important event in this regard [2] although other measures, including the first orbital spaceflight (1957) and release of animals to space, were done earlier [5].

Nowadays in medicine and biology, Yuri Gagarin is a gene in *Drosophila* which has codes for three proteins [14].

In the USA, the National Advisory Committee for Aeronautics (NACA) became NASA on 1958 [1].

The first human who landed and walked on the Moon on July 20, 1969, was Neil Armstrong (1930–2012, American astronaut and commander of the Apollo 11). His colleagues were Michael Collins and Edwin Aldrin [1].

The deployment of ISS in the low Earth orbit (LEO) was carried out to rotate the human body around the Earth at the height of about 400 km; then another great mission was performed. Ninety percent of gravity at ground level will be endured by astronauts in LEO [9].

So far, about 90 astronauts have arrived at the ISS, with an average length of stay of about 6 months (6). The Apollo program included human missions and spaceflights above LEO. The program ended in 1972. During this program, 12 astronauts traveled to the Moon, which lasted for a few days [10].

The Chinese are trying hard to build their space station which will be built around 2020 [11].

4. Spaceflights, Earth, Mars, and again Earth

Astronauts are preparing to travel to Mars now. Two and a half years are considered for this mission, part of it for spaceflight and the remaining for settlement and scheduled activities on the planet [15].

They are exposed to weightlessness in their 6-month course, which seems to make them physically very disable and inefficient at the end of the flight [7].

Most astronauts will encounter problems in spatial orientation and balance during the first few days after landing. They will be at risk for fractures and muscle tears during recovery period [9].

If the astronauts reach Mars with a weak and inadequate physical condition, having no solution to perform their basic tasks, the mission will face serious risks and will almost fail. Astronauts will endanger greater risks by sensory-motor impairment during spaceship control, off-vessel operations, or remote-controlled tasks. As long as the problems caused by weightlessness were not solved successfully, it is not wise to consider the long-term human journeys and missions like going to Mars [9].

Nowadays, human exploration missions to the Moon or Mars are considered as the next logical steps in the space era. In some cases, even human dispatch and migration, and colonization in other worlds, have also been announced. Almost all major national and international organizations in the world, as well as private investors and business plans, are currently developing roadmaps and related technologies to bring a healthy human to other planets of the solar system. Gradually, it seems that mankind is going to imagine about frequent and extended space trips, so that interplanetary traveling is equivalent to the concept of intercontinental flights. The human spaceflight perspective looks at the following to achieve in a short period:

- A significant number of travelers
- Longer flights and farther distances
- Sustainable deployment of humans to other planets as well as colonization

Many different challenges would be encountered that scientists and also politicians should overcome [9].

Most of the medical challenges that were faced during humanitarian missions and related steps were mainly due to radiation and effects of microgravity (or hypogravity) as well as psychological issues [8]. In future exploratory missions for the Moon and Mars, the crew will be exposed to problems of the permanent base establishment on the planet, long-duration flights and missions, radiation intensity, microgravity, and the different impressions of constraints and isolation. It poses several health issues that may be a limiting factor during these missions. The health and performance of the crew should be ensured during journeys on the transfer and discovery of the planet's surface, external vehicle activities, and after returning to the Earth. In particular, the mission to Mars includes more challenges: the planet's distance, travel time (at least 500 days), and the impossibility of suspending the missions that necessitates an entirely stable mission. There could not be any support from the Earth in the major health problem or technical one [9].

5. Types of simulation of space travel

1. Human:

   a. Bed rest (prolonged bed rest protocols and head-down tilt experiments, about 6°)

   Head tilt is the most used space analog in scientific researches. The head is tilted down (−6° almost in all cases) during lying in a bed for different periods according to the study. There would be a cephalic fluid shift, immobilization, and isolation in this kind of modeling [16].

   b. Separation and isolation environments (similar to studies conducted in Antarctica or research that people live in and operate in specific environments and chambers, such as those conducted by volunteers traveling to Mars) [17]

   c. Parabolic flight: although weightlessness will be executed only for some seconds [18].

   An airplane has alternative flies so that there would be three phases (normal gravity, hypergravity, and microgravity) in each period. Microgravity lasts about

20 seconds. Parabolic flight is the only simulation model that the person feels real microgravity. The airplane might fly until 30 periods maximally in a flight [18].

2. Computer simulation (especially in musculoskeletal biomechanics):

The data which are gathered or extracted from spaceflight and other experiments such as parabolic flights have been documented to validate the results of simulations [19].

3. Animal: Particularly about the points that could not be implemented even by human studies, such as:

   a. Bone and muscle radiation

   b. Detection of genetic effects on osteoporosis and muscle atrophy

   c. To identify the process of bone fracture restoration in the space travel environment (i.e., weightlessness and immune system alterations)

The main reason for performing animal models and simulation is that spaceflights are rare and expensive [20].

Several studies have been performed using tail suspension model, hind-limb suspension model, and so on (especially in rodents such as rat and mouse) [21, 22].

4. Cell [23]:

   a. Rotating wall vessel (a kind of two-dimensional clinostat)

   b. Random positioning machine (a kind of three-dimensional clinostat)

   c. Other types of clinostats

Scientists use ground-based models because of two reasons [24]:

- They have more control over the research situation.
- The cost would be much less.

Parabolic flight is such an analog. An airplane repeats a particular flight, so there would be a few seconds of free fall (weightlessness or no gravity) in each period.

Importance of human and animal simulations is stated as [5]:

- To study on high sample numbers
- Understanding human sexual differences (effectiveness of interaction)
- Reaction and interaction of different countermeasures (including nutrition, exercise, drugs, sleep, and others)

The reasons for not replacing cellular simulators instead of real flights are as the following [23]:

Cell simulators are important regarding creating insight into the suitability of future space missions, but it can be argued that these devices may still not replace the experiences of space travel. In this regard, two reasons can be counted:

1. Space travel has its complexities and complications since microgravity is not considered as the unique problem alone, but hypergravity, radiation, and vibration are also entangled.

2. The rotation kinetics of this simulator may cause fluid movement and shake in the chamber, which adds extra force to the cells present in the medium.

Therefore, it is necessary to consider that the results of these devices and techniques should be confirmed with the experiences in real microgravity.

6. The effects of space and its physiology on the human body

Astronauts are healthy individuals who will pass many courses to be well trained and finally highly selected. They should encounter different physiology until accommodation to pathophysiological alterations as soon as possible. Space medicine has some similarities (hypoxia, dysbarism, thermal support, acceleration, and response to high altitudes) with both aviation medicine and diving medicine [9]. These branches of medicine are taught separately, but there are aerospace medicine residency programs in a few countries (aviation medicine plus space medicine). There is only one combined aerospace and subaquatic medicine residency course and faculty in Iran.

Living and working on the Earth differ with space as mentioned earlier. Standing or sitting is the ideal position for doing the tasks and responsibilities. Gravity direction is perpendicular to the surface of our planet, so the body fluids are pulled down. The same occurs to our bones and muscles. The following paragraphs show the alterations of the human body during spaceflights briefly.

Some changes due to of microgravity exposure would become evident after a few days (i.e., motion sickness), some after weeks (i.e., cardiovascular deconditioning), and some even after months (i.e., osteoporosis and muscular atrophy) [9].

6.1 Immune system

Immune problems that have been identified and reported during spaceflights are:

- Alteration in adaptive immunity regulation
- Disordered interaction between adaptive and innate immunity
- Changes in the distribution of peripheral leukocytes
- Decrease in function of some leukocyte types
- Change in cytokine profiles
- More infectious disease incidence

- More allergic symptoms (even sometimes prolonged integumentary hypersensitivity reactions) [25]
- Weak immune reactions
- Possibility of increasing opportunistic infections and immunity disorders
- Increase in virulence of microorganisms [26]
- Hindering of thymopoiesis significantly [27]
- More latent viral reactivation [28]

There are more reports about immune system alterations [25, 26, 29], but some authors have not reported significant changes [30].

6.2 Muscular system

The most frequent change is atrophy of skeletal muscles. Alterations are described as:

- Atrophy and loss of function in lower limb muscles (especially the triceps muscle) [31]
- Decrease in performance of lower extremities muscles [32]
- Atrophy of muscle due to stem cell pool decrease [33]
- Spinal deconditioning (including atrophy of the paraspinal muscle group, reduction in curvatures of the spine) [34, 35]
- More damage in plantar flexors than extensors due to a decrease of fiber size and protein synthesis [9]
- Reduction of volume and strength in lower limb muscles [36]
- Significant loss in total body nitrogen and muscle volume in the Mir station [37]
- Muscle atrophy greater than 20 days of bed rest [38]

6.3 Cardiovascular system

These changes have been observed:

- Alterations in shape and mass of the heart, cardiac function changes, and arrhythmias [39]
- Redistribution of venous blood volume [40]
- Cephalad fluid shift, not any change in heart rate, a decrease in parasympathetic activity, an increase in output and stroke volume, and no change in sympathetic nerve activity [41]
- Loss of arterial pressure gradient [9]

- Changes in venous function [42]
- More sensitive to space radiation (opposite to previous studies) [12]

The heart does not work against gravity, so it has less work (with associated advantages and disadvantages) [9].

6.4 Respiratory system

Different results have been published by studies, so the comparison would not be easy due to the environmental bias and the variability of experimental protocols [9].

6.5 Blood

- Spaceflight anemia (reported frequently) [43, 44]
- Sustained increase of RBC and platelet [45]

6.6 Central nervous system (CNS)

Currently, there is little known about CNS alterations in space [46, 47]. The following structures are more affected than other parts of the CNS: cerebellar, sensorimotor, and vestibular brain regions [48].

6.7 Autonomic nervous system

Orthostatic stress has been reported, but its severity is less than on the Earth [49].

6.8 Sleep

Sleep problem is a common issue in space. It may affect the health of astronauts, so the safety of the mission would be endangered. Six causes for sleep alterations might be (1) factors related to cabin environment, (2) schedule adjustment, (3) impaired sleep discipline, (4) disordered circadian rhythm, (5) unreasonable work arrangement, and (6) mental and/or physical factors [50].

6.9 Eye

Spaceflight-associated neuro-ocular syndrome has been reported which is associated with optic disk edema (unilateral and bilateral), flattening of the eyeball, folds of choroid and retina, hyperopic refractive error shifts, and nerve fiber layer infarcts [51].

6.10 Skeletal system

Most changes occur in the skeletal system. Osteoporosis can be a major obstacle to the actions and activities of astronauts, especially in long-term missions and/or traveling to other planets [52].

Skeletal changes in spaceflights seem to be the most important alterations because they may limit or even stop a space program [53]. Rate of monthly 1–2% bone loss has been postulated during spaceflights [52] and 0.5–1.5% by some other investigators [54]. It could be up to 15%. Returning to normal values after landing on the Earth is much slower, maybe only 6% annually [55]. Vico et al. did not observe recovery at skeletal sites the year after return from ISS in 2017 [56]. In 2010, Dana et al. stated prolonged consequences of bone loss even after 2.5 years following a stay on ISS [57].

It has been shown that bones of the lower extremities become more osteoporotic in comparison to other parts of the body. Skull bones are less affected, and even some report increased density of the skull. Then we may assume the most critical factor of osteoporosis to be unloading [16].

Several articles have shown the skeletal system and its physiology alterations due to spaceflight [54, 56–60]. LeBlanc et al. reported a 5% decrease of bone mineral density was observed in 92% of 60 American and Russian astronauts during long-term flights lasting 4–6.5 months [61]. The review by LeBlanc et al. showed that weight-bearing bones had more bone mineral loss [62]. The results of Skylab were repeated in the Soyuz and Mir missions [9]. 1,25-Vitamin D and its precursor reduced significantly in the Mir 18 mission [9]. Lang et al. claimed their data shows that full recovery of bone density was not complete even after 1 year [59]. Experiments conducted in the Mir station demonstrated 13% of osteoporosis in the pelvis which had the most significant range of bone loss in comparison with other parts of the skeleton [37].

7. Vitamin D

In missions and space travels, the health of the astronauts and, consequently, its maintenance are essential, and it would be vital in some cases and situations. Proper nutrition will positively affect human health; vitamins such as vitamin D have important roles too.

Leach and Rambaut's research on Skylab 4, published in 1977, showed that the amount of vitamin D was lower in astronauts, even with supplementation of 500 units per day. Their mission was 84 days long, but such results were not observed for previous Skylab's short-term missions (28 and 59 days) [63].

The studies on astronauts of the Mir space station showed that the amount of vitamin D had dropped. The reasons were considered as the following [52]:

- Consumption of low vitamin
- Lack of exposure to ultraviolet radiation

In 2001, Smith and colleagues conducted a study on nutrition status assessment in 2 isolated environments (2 missions 60 and 91 days long, 4 participants, on the Earth) and a 4-month residence of two astronauts at the Mir station. They found that the amount of vitamin D was 32–36% of the average value [64].

Smith et al. published the results of nutritional status of 11 astronauts with long-term missions on ISS (128–195 days, 2000–2004). The program contained 10 μg/d of supplementary vitamin D; 10 micrograms of cholecalciferol was in each supplement. A significant decrease (25%) was observed on 25-hydroxycholecalciferol after the mission, and bone markers showed increased resorption [63].

The report indicates weekly consumption of 5.7 ± 4.0 supplementary vitamin D per (and mean of 3.5 ± 2.9 supplementary multivitamins) [63]. It does not seem to be a sufficient amount of vitamin D for compensation.

Following the launch of the ISS, the order for supplementing vitamin D was issued to astronauts. Meanwhile, the twenty-first century had come. In the first decade of this century, vitamin D was considered by anyone, and a wave of research and reports began. Everywhere we heard and read about this vitamin, from newspapers, magazines, radios, and TVs. The Internet was also overwhelming. Are considerable number of articles and papers, and surprising notes showing that vitamin D is of tremendous importance and everything from cancer and a variety of serious diseases and dangers to a variety of even chronic infections such as tuberculosis would be impacted. The results of the mentioned wave were expanded from the Earth to the sky and space. Fortunately, this was a positive aspect of the case, so the daily dosage of vitamin D increased from 400 to 800 units for astronauts on ISS [65].

The research published by Smith et al. in 2012 was based on 13 astronauts (9 males and 4 females, between 2006 and 2009, length of mission as a long one between 4 and 6 months). It was the first study of this type that showed the mineral density of bones, and risk of osteoporosis would be reduced in long-term space mission by severe exercise and adequate feeding (good energy, plus 800 units of vitamin D per day, which was twice the amount of vitamin D prescribed in previous missions). Of course, the results were achieved by bone remodeling [66] more than by bone resorption reduction. Blood and urine specimens were prepared before, during, and after space travel, but densitometry was performed before and after the trip. Their findings can help ensure that vitamin D is sufficient to live in an environment that does not have any exposure to ultraviolet radiation and with limited food available (as the source of vitamins).

7.1 Vitamin D and the positive point

Vitamin D is easy to use and available. Its cost is not so much, and it may be prescribed as an efficient strategy in hindrance and/or prevention of space-induced changes (particularly in bone) [21].

7.2 Vitamin D and the negative point

An important subject that should be considered is vitamin D does not have an infinite capacity for preventing osteoporosis (cellular process) [22].

7.3 Vitamin D story continued

In another research, our team measured the bone mineral density of 14 rats' femur and demonstrated morphologic changes for the contralateral femur after supplemental calcium/vitamin D. The bone mineral density and bone mineral content had a significant increase in the experiment group. The outer cortical bone thickness was also higher [21].

Astronauts should be healthy and able to maintain their health in order to be capable of performing their tasks as the best as possible. It should be implemented during and even after space travel (either landing on Mars or returning to the Earth). In this case, nutrition is of particular importance for the body as a whole and cellular processes. The food is packaged and ready for astronauts, which should be of good quality and, at the same time, meet the requirements of scientific references [67].

Cooper et al. published their research results on astronauts' food quality assessment in 2017 [67]. They examined 24 micronutrients including vitamin D in 109 food packages for 3 years and determined the amount of vitamin D at the beginning of the study, 1 and 3 years later. It turned out that vitamin D is not even adequate at the beginning of the study compared to standard food tables! It was less than 50% of the required amount!

Please read the last paragraph again. It is amazing. Astronauts are selected and qualified persons after many tests and exams. It is rational that their food should be as suitable as possible and highly selected. The scientists had noticed the decrease of vitamin D in several studies, so they proposed the causes and solutions, but it seems that nobody noticed and tested the real amount of vitamin in foods (of course before Cooper et al. [67]). Modern life has changed our lifestyle; we become obese more and more, limited body activities and exercises, not going out of the buildings, using more sunscreens, maybe more makeups (especially in third world countries), covering all of the body (religious or national belief). Sunlight does not touch our body and skin. Vitamin D would not be produced. Then we should rely on the foodstuff, but they do not have sufficient vitamin D; a great disaster begins! It is so simple!

Therefore, the first solution would be the necessity of providing vitamin supplement for astronauts [67]. Since 2006, it has been announced that vitamin D supplementation would be part of the regular nutrition program in spaceflights [65].

Other solutions to this problem include:

- The use of vitamin D-rich or fortified foods in the diet of astronauts
- Specific/specialized techniques for storage and maintaining adequate amounts of vitamin D in foods during travel

In 2016, Wotring examined nine drugs which had been returned from ISS for the sake of their chemical power and degradation percentage. Their maintenance under the unusual situation in space travel may be associated with increased drug decomposition. Until now, the drugs are replaced by new ones before the expiration date. It is unlikely to be possible to replace expired drugs on a long journey such as to Mars. Wotring failed to provide a complete guide to this issue because there was no control sample on the Earth and proposed more studies should be carried out [68].

Although prescribing vitamin D supplements can be proposed as a nutritional countermeasure, supplements may also degrade and lose their effect. In this regard, comprehensive research should be provided [67].

Zwart and colleagues published their research results in 2009, considering that maintaining and supplying food in spaceships are vital and essential to the success of the mission and health of the astronauts. Based upon past studies conducted on the Earth, they announced that long-term storage of vitamins or exposure to radiation might alter some of these, so that their intake does not affect, or even dangerous and harmful. They sent four identical kits containing foods and vitamins (including vitamin D supplements) to ISS in 2006. They also kept a control group (similar packages) at ground level. The experiments were carried out on four occasions (on the days 13th, 353rd, 596th, and 880th). The amount of vitamin remaining after 596 days changed due to the length of storage time in most foods, but the final results showed that there was no difference between space travels and the Earth's surface [69].

Some studies (including Smith et al.) have suggested that the amount of vitamin D in long-term space travel would change, with possible causes as [70]:

1. Metabolism in various stages of vitamin D metabolism:

    a. Absorption

b. Consumption

c. Excretion

2. Food decomposition due to:

a. Time spent

b. Other unknown causes

Smith et al. have referred to vitamin D reduction in 3- to 6-month space travel (with unknown mechanisms) in three different studies (published in 1999 and 2 papers in 2005). This was while the spaceship crew had consumed the supplement too. No mechanism is known for such a reduction, although two factors can be considered alone or in combination [69].

- Long-term food storage
- Ionizing radiation

These two factors can disrupt the feeding of astronauts through the following routes [70]:

- Food decomposition
- Reduced bioavailability
- Food oxidation and the emergence of inappropriate odors

Another point to note is the effect of radiation on vitamins. Vitamins are considered to be human-protecting sources that do not have any substitute. If we consider vitamins to be isolated from food, then radiation would have little effect on them, because they are considered as small molecules. However, if the radiation hits the vitamins inside the food (where molecules of water and lipids are around them), there would be radicals that might affect vitamins negatively [69].

Better than any conclusion, please go straight to the summary and find it there.

8. Summary

Human chose spaceflights many years ago. The space environment has its distinct physiology which is different from the Earth. Dramatic changes would occur in the human body totally and in the skeletal system as a part. There would be no trivial sun or UV exposure in spaceships; so everything is ready for the advent of osteoporosis. In such situations, vitamin D had come to the scene, to corporate positive effects in the skeleton and other body systems. Studies have shown some promising results, but as revealed earlier, the main concern and problem would be the amount of vitamin D in prepared foods for astronauts (lesser than expected and standard values). It seems that distinguished scientists have noticed the subject and have tried some solutions, maybe a process, suitable even on the Earth.

Acknowledgements

The author wishes to thank for no help or assistance of Aerospace and Subaquatic Medicine Faculty staff, especially Dr. Reza Eslami and Dr. Abolfazl Khademi.

Conflict of interest

There is no conflict of interest.

Author details

Amir Khoshvaghti
Infectious Diseases Research Center, Aerospace and Subaquatic Medicine Faculty, AJA University of Medical Sciences, Tehran, Iran

*Address all correspondence to: anatomygray2009@gmail.com

References

[1] Launius RD. Neil Armstrong (1930-2012). Nature. 2012;**489**(7416): 368

[2] West JB. Historical aspects of the early Soviet/Russian manned space program. Journal of Applied Physiology (1985). 2001;**91**(4):1501-1511

[3] Harsch V, Bardrum B, Illig P. Lilienthal's fatal glider crash in 1896: Evidence regarding the cause of death. Aviation, Space, and Environmental Medicine. 2008;**79**(10):993-994

[4] Schlenoff DC. The equivocal success of the Wright brothers. Scientific American. 2003;**289**(6):94-97

[5] Morey-Holton ER, Hill EL, Souza KA. Animals and spaceflight: From survival to understanding. Journal of Musculoskeletal & Neuronal Interactions. 2007;**7**(1):17-25

[6] Chapin SL. Aviationists 200 years ago: The montgolfier brothers and the invention of aviation, 1783-1784. Science. 1983;**221**(4610):542-543

[7] Vandenbrink JP, Kiss JZ. Space, the final frontier: A critical review of recent experiments performed in microgravity. Plant Science. 2016;**243**:115-119

[8] Thirsk R et al. The space-flight environment: The International Space Station and beyond. CMAJ. 2009;**180**(12):1216-1220

[9] Demontis GC et al. Human pathophysiological adaptations to the space environment. Frontiers in Physiology. 2017;**8**:547

[10] Lacquaniti F et al. Human locomotion in hypogravity: From basic research to clinical applications. Frontiers in Physiology. 2017;**8**:893

[11] Wang L, Li Z, Tan C, Liu S, Zhang J, He S, Zou P, Liu W, Li Y. Physiological effects of weightlessness: Countermeasure system development for a long-term Chinese manned spaceflight. Frontiers in Medicine. 2018 Apr 25. DOI: 10.1007/s11684-017-0587-7. [Epub ahead of print] Review. PubMed PMID: 29693211

[12] Boerma M et al. Space radiation and cardiovascular disease risk. World Journal of Cardiology. 2015;**7**(12):882-888

[13] Chancellor JC et al. Limitations in predicting the space radiation health risk for exploration astronauts. npj Microgravity. 2018;**4**:8

[14] Texada MJ et al. Tropomyosin is an interaction partner of the Drosophila coiled coil protein yuri gagarin. Experimental Cell Research. 2011;**317**(4):474-487

[15] Cooper M, Douglas G, Perchonok M. Developing the NASA food system for long-duration missions. Journal of Food Science. 2011;**76**(2):R40-R48

[16] Hargens AR, Vico L. Long-duration bed rest as an analog to microgravity. Journal of Applied Physiology (1985). 2016;**120**(8):891-903

[17] Pagel JI, Chouker A. Effects of isolation and confinement on humans-implications for manned space explorations. Journal of Applied Physiology (1985). 2016;**120**(12):1449-1457

[18] Shelhamer M. Parabolic flight as a spaceflight analog. Journal of Applied Physiology (1985). 2016;**120**(12):1442-1448

[19] Gerber B et al. A computer simulation of short-term adaptations of cardiovascular hemodynamics in

microgravity. Computers in Biology and Medicine. 2018;**102**:86-94

[20] Globus RK, Morey-Holton E. Hindlimb unloading: Rodent analog for microgravity. Journal of Applied Physiology (1985). 2016;**120**(10):1196-1206

[21] Kouhnavard M et al. Effects of vitamin D and calcium supplementation on micro-architectural and densitometric changes of rat femur in a microgravity simulator model. Iranian Red Crescent Medical Journal. 2014;**16**(6):e18026

[22] Hashemian SJ et al. Effect of calcitriol supplementation and tail suspension on serum biomarkers of bone formation in rats. Journal of Diabetes and Metabolic Disorders. 2015;**14**:14

[23] Li N et al. Microgravity-induced alterations of inflammation-related mechanotransduction in endothelial cells on board SJ-10 Satellite. Frontiers in Physiology. 2018;**9**:1025

[24] Ramachandran V et al. Effects of spaceflight on cartilage: Implications on spinal physiology. Journal of Spine Surgery. 2018;**4**(2):433-445

[25] Crucian BE et al. Immune system dysregulation following short- vs long-duration spaceflight. Aviation, Space, and Environmental Medicine. 2008;**79**(9):835-843

[26] Taylor K et al. Toll mediated infection response is altered by gravity and spaceflight in Drosophila. PLoS One. 2014;**9**(1):e86485

[27] Benjamin CL et al. Decreases in thymopoiesis of astronauts returning from space flight. JCI Insight. 2016;**1**(12):e88787

[28] Mehta SK et al. Latent virus reactivation in astronauts on the international space station. npj Microgravity. 2017;**3**:11

[29] Crucian B et al. Alterations in adaptive immunity persist during long-duration spaceflight. npj Microgravity. 2015;**1**:15013

[30] Crucian BE et al. Immune status, latent viral reactivation, and stress during long-duration head-down bed rest. Aviation, Space, and Environmental Medicine. 2009;**80**(5 Suppl):A37-A44

[31] Rittweger J et al. Sarcolab pilot study into skeletal muscle's adaptation to long-term spaceflight. npj Microgravity. 2018;**4**:18

[32] Mulavara AP et al. Physiological and functional alterations after spaceflight and bed rest. Medicine and Science in Sports and Exercise. 2018;**50**(9):1961-1980

[33] Hosoyama T et al. Microgravity influences maintenance of the human muscle stem/progenitor cell pool. Biochemical and Biophysical Research Communications. 2017;**493**(2):998-1003

[34] Holt JA et al. WISE 2005: Aerobic and resistive countermeasures prevent paraspinal muscle deconditioning during 60-day bed rest in women. Journal of Applied Physiology (1985). 2016;**120**(10):1215-1222

[35] Adams GR, Caiozzo VJ, Baldwin KM. Skeletal muscle unweighting: Spaceflight and ground-based models. Journal of Applied Physiology (1985). 2003;**95**(6):2185-2201

[36] Gopalakrishnan R et al. Muscle volume, strength, endurance, and exercise loads during 6-month missions in space. Aviation, Space, and Environmental Medicine. 2010;**81**(2):91-102

[37] LeBlanc A et al. Muscle volume, MRI relaxation times (T2), and

body composition after spaceflight. Journal of Applied Physiology (1985). 2000;**89**(6):2158-2164

[38] Akima H et al. Effect of short-duration spaceflight on thigh and leg muscle volume. Medicine and Science in Sports and Exercise. 2000;**32**(10):1743-1747

[39] Zhu H, Wang H, Liu Z. Effects of real and simulated weightlessness on the cardiac and peripheral vascular functions of humans: A review. International Journal of Occupational Medicine and Environmental Health. 2015;**28**(5):793-802

[40] Taibi A et al. Investigation of cerebral venous outflow in microgravity. Physiological Measurement. 2017;**38**(11):1939-1952

[41] Otsuka K et al. Long-term exposure to space's microgravity alters the time structure of heart rate variability of astronauts. Heliyon. 2016;**2**(12):e00211

[42] Fortrat JO et al. Altered venous function during long-duration spaceflights. Frontiers in Physiology. 2017;**8**:694

[43] Tavassoli M. Anemia of spaceflight. Blood. 1982;**60**(5):1059-1067

[44] Smith SM. Red blood cell and iron metabolism during space flight. Nutrition. 2002;**18**(10):864-866

[45] Kunz H et al. Alterations in hematologic indices during long-duration spaceflight. BMC Hematology. 2017;**17**:12

[46] Jandial R et al. Space-brain: The negative effects of space exposure on the central nervous system. Surgical Neurology International. 2018;**9**:9

[47] Clement G, Ngo-Anh JT. Space physiology II: Adaptation of the central nervous system to space flight—Past, current, and future studies. European Journal of Applied Physiology. 2013;**113**(7):1655-1672

[48] Van Ombergen A et al. The effect of spaceflight and microgravity on the human brain. Journal of Neurology. 2017;**264**(Suppl 1):18-22

[49] Mandsager KT, Robertson D, Diedrich A. The function of the autonomic nervous system during spaceflight. Clinical Autonomic Research. 2015;**25**(3):141-151

[50] Wu B et al. On-orbit sleep problems of astronauts and countermeasures. Military Medical Research. 2018;**5**(1):17

[51] Lee AG et al. Space flight-associated neuro-ocular syndrome (SANS). Eye (London, England). 2018;**32**(7):1164-1167

[52] Smith SM et al. Fifty years of human space travel: Implications for bone and calcium research. Annual Review of Nutrition. 2014;**34**:377-400

[53] Vico L, Hargens A. Skeletal changes during and after spaceflight. Nature Reviews Rheumatology. 2018;**14**(4):229-245

[54] Lang T et al. Cortical and trabecular bone mineral loss from the spine and hip in long-duration spaceflight. Journal of Bone and Mineral Research. 2004;**19**(6):1006-1012

[55] Grimm D et al. The impact of microgravity on bone in humans. Bone. 2016;**87**:44-56

[56] Vico L et al. Cortical and trabecular bone microstructure did not recover at weight-bearing skeletal sites and progressively deteriorated at non-weight-bearing sites during the year following International Space Station missions. Journal of Bone and Mineral Research. 2017;**32**(10):2010-2021

[57] Dana Carpenter R et al. Long-term changes in the density and structure of the human hip and spine after long-duration spaceflight. Acta Astronautica. 2010;**67**(1):71-81

[58] Sibonga JD et al. Evaluating bone loss in ISS astronauts. Aerospace Medicine and Human Performance. 2015;**86**(12 Suppl):A38-a44

[59] Lang TF et al. Adaptation of the proximal femur to skeletal reloading after long-duration spaceflight. Journal of Bone and Mineral Research. 2006;**21**(8):1224-1230

[60] Sibonga JD et al. Recovery of spaceflight-induced bone loss: Bone mineral density after long-duration missions as fitted with an exponential function. Bone. 2007;**41**(6):973-978

[61] LeBlanc A et al. Bone mineral and lean tissue loss after long duration space flight. Journal of Musculoskeletal & Neuronal Interactions. 2000;**1**(2):157-160

[62] LeBlanc AD et al. Skeletal responses to space flight and the bed rest analog: A review. Journal of Musculoskeletal & Neuronal Interactions. 2007;**7**(1):33-47

[63] Smith SM et al. The nutritional status of astronauts is altered after long-term space flight aboard the International Space Station. The Journal of Nutrition. 2005;**135**(3):437-443

[64] Smith SM et al. Nutritional status assessment in semiclosed environments: Ground-based and space flight studies in humans. The Journal of Nutrition. 2001;**131**(7):2053-2061

[65] Lang T et al. Towards human exploration of space: The THESEUS review series on muscle and bone research priorities. npj Microgravity. 2017;**3**:8

[66] Smith SM et al. Benefits for bone from resistance exercise and nutrition in long-duration spaceflight: Evidence from biochemistry and densitometry. Journal of Bone and Mineral Research. 2012;**27**(9):1896-1906

[67] Cooper M, Perchonok M, Douglas GL. Initial assessment of the nutritional quality of the space food system over three years of ambient storage. npj Microgravity. 2017;**3**:17

[68] Wotring VE. Chemical potency and degradation products of medications stored over 550 Earth days at the International Space Station. The AAPS Journal. 2016;**18**(1):210-216

[69] Zwart SR et al. Assessment of nutrient stability in foods from the space food system after long-duration spaceflight on the ISS. Journal of Food Science. 2009;**74**(7):H209-H217

[70] Smith SM et al. Bone markers, calcium metabolism, and calcium kinetics during extended-duration space flight on the mir space station. Journal of Bone and Mineral Research. 2005;**20**(2):208-218

Section 4

Clinical Applications of Vitamin D Measurement and Replacement

Chapter 6

Vitamin D in Rheumatic Diseases: Interpretation and Significance

Binit Vaidya and Shweta Nakarmi

Abstract

The pleiotropic effects of vitamin D on the various metabolic, anticancer, and immunomodulatory functions of the body based on the presence of vitamin D receptors (VDR) on various cell types has been recognized worldwide now. Of few understood mechanisms of immunomodulatory actions of vitamin D are the suppressive action on the maturation of antigen-presenting cells and decrease in the levels of pro-inflammatory cytokines. Vitamin D deficiency has been implicated in the immune diseases like rheumatic diseases, asthma, psoriasis, and multiple sclerosis. Vitamin D deficiency has been associated with increased frequency and severity of disease flares in rheumatic diseases like lupus and rheumatoid arthritis. Other studies have shown higher prevalence of persistence and evolution in to more definite rheumatic disorder in undifferentiated arthritis and undifferentiated connective tissue disorder patients with vitamin D deficiency. Multiple factors like avoidance of sunlight, the use of corticosteroids and hydroxychloroquine, skin pigmentation, etc. should be considered when evaluating vitamin D levels in these patients, needless to say the consideration of higher-dose supplement for these patients. It is thus prudent that all patients with established or undifferentiated rheumatic diseases are evaluated for vitamin D status and an adequate supplementation is recommended to prevent the associated consequences.

Keywords: vitamin D, vitamin D deficiency, immunomodulation, rheumatic diseases, inflammation

1. Introduction

Vitamin D has two bioequivalent forms, D2 (ergocalciferol) and D3 (cholecalciferol). It is synthesized mainly from 7-dehydrocholesterol in keratinocytes of the skin stimulated by UVB of sunlight and metabolized in the liver to 25(OH)D and subsequently converted to its active form 1,25$(OH)_2$D in the kidney [1, 2]. It maintains calcium and phosphorus homeostasis, optimizes bone health and muscle function and immunomodulation, has antiproliferative effect on keratinocytes, and suppresses cytokine production [3, 4]. Serum 25(OH)D3 level of at least 50 nmol/l is considered to be optimal for bone health and extra skeletal effects [5]. The term hypovitaminosis D includes vitamin D insufficiency and deficiency. Vitamin D insufficiency is defined as a serum 25(OH)D concentration of 21–29 ng/ml (50–75 nmol/L), whereas deficiency means serum 25(OH)D level of <20 ng/ml (<50 nmol/L) [6].

IntechOpen

Vitamin D is also known as the sunshine vitamin. The importance of sunlight for human health came into light with the industrial revolution in Northern Europe [7]. Sniadecki first published an article in 1822 about high prevalence of rickets in children who lived in the inner city in comparison to those who lived in the rural areas [8]. Many observations regarding the sun exposure and rickets have been published in the course of time. Studies have also revealed the high prevalence of vitamin D deficiency in general population, mostly owing to lack of sun exposure.

2. Vitamin D structure, synthesis, and metabolism

Vitamin D is a fat-soluble seco-steroid made from four cholesterol rings IOM (Institute of Medicine) [9]. It has two bioequivalent forms, D2 (ergocalciferol) and D3 (cholecalciferol). Vitamin D2 is derived from the plant sterol ergosterol [1, 9]. Vitamin D from the diet or dermal synthesis is biologically inactive [10]. Vitamin D3 is synthesized mainly from 7-dehydrocholesterol in keratinocytes of the skin stimulated by UVB of sunlight [3]. Under the influence of sunlight (ultraviolet radiation, action spectrum 280–320 nM, or UVB), 7-dehydrocholesterol in the epidermis is converted to vitamin D (**Figure 1**). Keratinocytes express the vitamin D receptor (VDR) due to which they are capable of responding to the 1,25$(OH)_2$D produced [11]. Both UVB intensity and skin pigmentation level contribute to the rate of D3 formation [12]. D3 is converted to 25(OH)D (calcidiol) in the liver by a number of enzymes. 25-Hydroxyvitamins D2 and D3 produced by the liver enter the circulation and the kidney bound to vitamin D-binding protein. The kidney metabolizes 25(OH)D to the active metabolite 1,25$(OH)_2$D3.

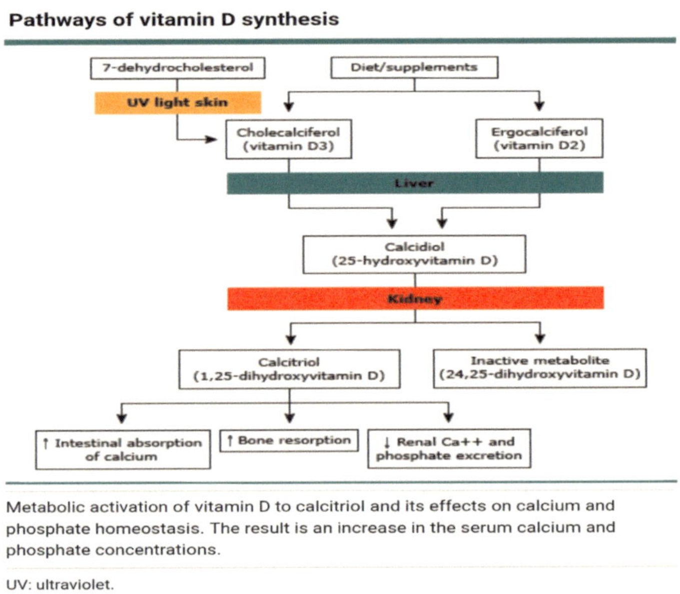

Figure 1.
Overview of vitamin D, UpToDate 2019.

3. Sources of vitamin D

More than 90% of vitamin D requirement comes from sunlight. According to Holick, exposure of ≥20% of the body's surface to either direct sunlight or tanning bed radiation is effective in increasing blood concentrations of vitamin D3 and 25-hydroxyvitamin D3 [25(OH)D3]. One minimal erythemal dose (MED) is equivalent to 10–50 times the recommended intakes [3]. Oily fishlike salmon, mackerel, sardines, and cod liver oil are also considered good sources of vitamin D [13].

4. Causes of vitamin D deficiency

There are numerous causes of vitamin D deficiency. Decreased synthesis from the skin is one of the main causes. Less exposure to sunlight, lifestyle, skin pigmentation, abundant use of sunscreen lotion, and geographical variation owe to less dermal synthesis. Decreased bioavailability due to malabsorption, decreased synthesis of active form of vitamin D due to liver or renal failure, or increased catabolism with the use of various medications like glucocorticoids and anticonvulsants also cause deficiency. VDD is also seen in diseases like rickets, osteomalacia, hyperparathyroidism, and granulomatous disorders [14].

5. Daily recommended dose of vitamin D

The daily recommended allowance of vitamin D is 400–600 IU in children and adults and 800 IU in adults >70 years [9].

6. Clinical applications

1. Bone and vitamin D: Vitamin D maintains calcium and phosphorus homeostasis and optimizes bone health and muscle function [4]. Adequate vitamin D is necessary to prevent rickets and osteomalacia [3]. Though still controversial, it is reported that calcium and vitamin D supplementation prevents fall risk and also decreases the osteoporotic fracture in older adults [15]. Vitamin D supplementation of 700–800 IU per day reduces falls and fractures in older adults [16].

2. PTH and vitamin D: There is an inverse relationship between circulating 25 (OH)D levels and parathyroid hormone (PTH) [1].

3. Muscle function and vitamin D: There are several studies that suggest a relationship between vitamin D and muscle function. Though improvement of muscle strength with supplementation of vitamin D has been observed in few trials [17, 18], the causal relation has not been established yet.

4. Skin and vitamin D: $1,25(OH)_2D$ analogs calcipotriol and maxacalcitol can be used for the treatment of the hyperproliferative skin diseases like psoriasis [3, 19] and non-melanoma skin cancer [20].

5. Cancer and vitamin D: Vitamin D deficiency has been associated with cancers, especially colorectal [21, 22]. Calcium and $1,25(OH)_2D3$ participate in the regulation of keratinocyte proliferation and differentiation and may prevent the development of skin cancer [20]. Observational studies have shown the relationship between vitamin D deficiency and carcinoma of the breast, colon, and thyroid [3], but the results are not consistent.

6. Immune system and vitamin D: Vitamin D is a potent immunomodulator. 1,25 $(OH)_2D$ decreases the maturation of dendritic cells (DCs) decreasing their ability to present antigen and to activate T cells [23]. Furthermore, it suppresses production of IL-12 (important for Th1 development), IL-23, and IL-6 (important for Th17 development) [24]. But, there is no any approved vitamin D drug for immune modulation [1]. Studies have suggested an association of vitamin D with autoimmune diseases like multiple sclerosis [25] and asthma [26].

7. Cardiovascular disease and vitamin D: There is an inverse relationship between vitamin D deficiency and risk of heart disease, myocardial infarction, and early death. Low vitamin D causes increased parathyroid hormone release, inflammation, proliferation of vascular smooth muscle cells, insulin resistance, thrombogenicity, dyslipidemia, and progressive extracellular matrix remodeling. All of these are associated with increased risk of ischemic heart disease, myocardial infarction, and early death [27, 28].

8. Diabetes mellitus and vitamin D: Vitamin D deficiency is associated with insulin resistance. 1,25(OH)2D promotes increased lipogenesis and decreased lipolysis. The pancreatic B cell expresses the VDR, and $1,25(OH)_2D$ promotes insulin secretion [29].

9. Neurological disorder and vitamin D: Vitamin D plays an important role in brain development as it has effects on neuronal proliferation, differentiation, migration, and apoptosis [30].

7. Vitamin D and immunomodulation

Vitamin D is involved in modulation of immune responses and has an important role in some autoimmune diseases like multiple sclerosis, diabetes mellitus, psoriasis, systemic lupus erythematosus (SLE), RA, etc. [31]. The biological effects or immunomodulation is mediated by the vitamin D receptor (VDR) which belongs to the nuclear hormone receptor family and is expressed in most cell types including macrophages, dendritic cells, B and T lymphocytes, and neutrophils [31, 32].

Along with the modulatory effects on T and B cell functions, VDR agonists inhibit the differentiation and maturation of DCs, thus influencing the function of DCs and promoting tolerogenic properties that favor the induction of regulatory T cells. VDR also downregulate expression of the costimulatory molecules CD40, CD80, and CD86, decrease production of IL-12, and increase production of IL-10. The inhibition of DC differentiation and maturation and production of pro-inflammatory mediators play an important role in the immunoregulatory activity of $1,25(OH)_2D3$ [33, 34].

$1,25(OH)_2D3$ also plays an important role in the maintenance of B cell homeostasis. It has potent effects on functions of B cell, including induction of apoptosis and inhibition of proliferation, generation of memory B cells, plasma cell differentiation, and immunoglobulin production [35].

According to Grant, there is evidence in support of vitamin D reducing the risk of many autoimmune diseases including such as multiple sclerosis and type 1 diabetes mellitus. However, evidence for rheumatoid arthritis, osteoarthritis, type 2 diabetes mellitus, hypertension, and stroke is weak [36].

8. SLE

Vitamin D deficiency is quite prevalent in SLE patients which may be attributable to various reasons. Avoidance of sunshine, photoprotection, renal insufficiency, and the use of medications which alter the metabolism of vitamin D or downregulate the functions of the vitamin D receptor like glucocorticoids, anticonvulsants, antimalarials, and the calcineurin inhibitors are some of the causes of VDD as shown in **Figure 2** [37]. In a study by Toloza, vitamin D insufficiency was found in 66.7% and deficiency in 17.9% of SLE patients [38]. The frequency varied in different studies: Saudi Arabia (89.7%) [39], Norway (82%) [40], Poland (71%) [41], Hong Kong (27%) [42], and the United States (20%) [43]. Low serum vitamin D levels were related to cumulative glucocorticoid dose [38]. Corticosteroids accelerate the catabolism of 25(OH)D and 1,25$(OH)_2$D and have a significant role in secondary osteoporosis [44]. Patients taking corticosteroids often require higher daily doses of vitamin D to maintain adequate levels [45]. Similarly, a commonly used antimalarial, hydroxychloroquine (HCQ), inhibits conversion of 25(OH)D to 1,25$(OH)_2$D leading to low levels of vitamin D [46].

A review by Sakthiswary demonstrated a substantial evidence in support of the association between vitamin D levels and SLE disease activity. However, vitamin D level is not associated with organ damage [47]. A study by Suzan showed a significant negative correlation that existed between 25(OH)D and anti-dsDNA and a positive correlation between 25(OH)D levels and C4 [48]. Another similar study showed a significant negative correlation between the serum concentration of vitamin D and the standardized values of disease activity scores as measured by the SLEDAI-2K and ECLAM scales [49]. An Australian study showed that low vitamin D

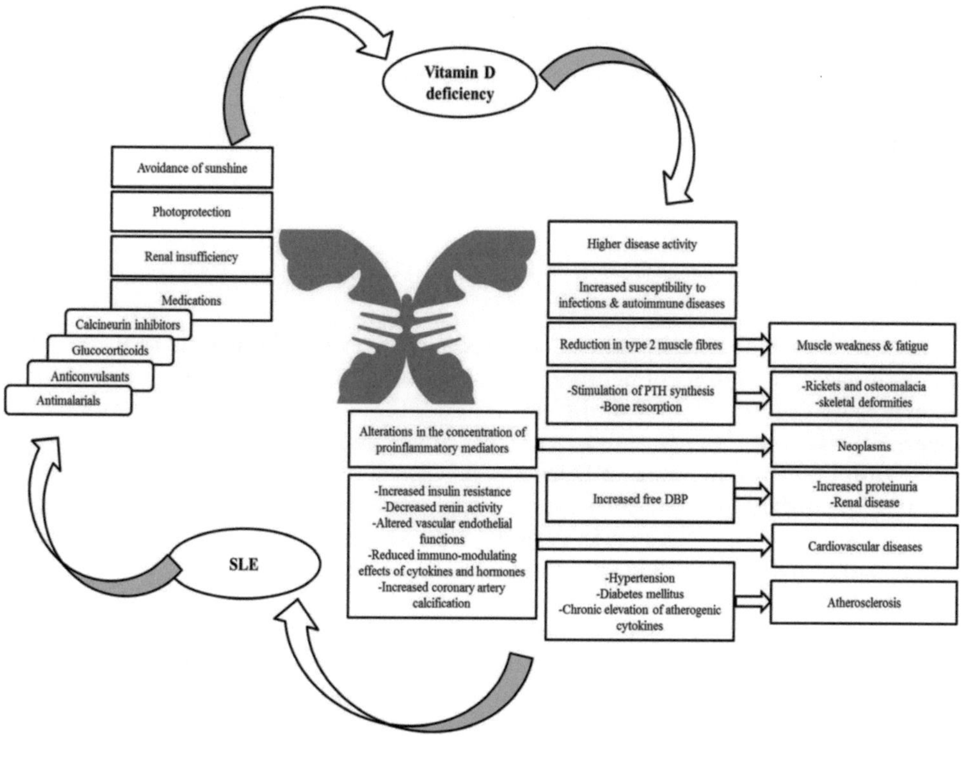

Figure 2.
The two-sided relation between vitamin D and SLE showing that low levels of vitamin D resulted from SLE and SLE complications that come from vitamin D deficiency [37].

was associated with a higher disease activity and an increase in serum vitamin D was associated with reduced disease activity over time [50]. Improving vitamin D status may improve other common manifestations as well, such as fatigue [51] and cognitive dysfunction [52].

9. RA

The role of hypovitaminosis D in the pathogenesis of rheumatoid arthritis has been the topic of interest in the recent past. Lower vitamin D levels possess increased risk for RA [53]. 1,25$(OH)_2$D3 contributes to the regulation of matrix metalloproteinase and prostaglandin E2 production by synovial fibroblasts and articular chondrocytes in RA [54].

High rates of vitamin D deficiency have been observed in patients with rheumatic diseases. A study demonstrated that in patients with RA, VDD was seen in 64% and insufficiency in 28%. Similarly, in spondyloarthritis (SpA) patients 48% had VDD and 35% had insufficiency [55]. The prevalence of VDD is quite in RA patients. The COMEDRA study showed that 55.8% of RA patients had vitamin D insufficiency and 3.6% had deficiency [56]. Eighty-four percent of RA patients were VDD in a recently published study by Meena. It also showed a significant inverse correlation between serum vitamin D levels and RA disease activity [57]. A meta-analysis by Lee and Bae supported this result suggesting that the vitamin D level is associated with susceptibility to RA and RA activity [58]. Similarly, a negative association between serum vitamin D and RA disease activity was demonstrated in few studies [59–62]. Levels of 25(OH)D3 were also found to be negatively correlated to CRP and ESR [62]. However, relationship between 25(OH)D and levels of rheumatoid factor or anti-cyclic citrullinated peptide antibodies has not been established yet [63].

The COMORA study showed that vitamin D was insufficient in 54.6% and deficient in 8.5% of the RA patients. Low levels of vitamin D were associated with disease activity of RA and corticosteroid dosage and comorbidities like lung disease and osteoporosis therapy [64].

10. CTD

In comparison to healthy adults, VDD is more prevalent in people with autoimmune diseases including connective tissue diseases (CTDs) [65]. It may also have a pivotal role in progression of undifferentiated CTDs to well-defined and more severe disease [65]. There are few evidences which showed the antifibrotic property of vitamin D [66]. Low vitamin D levels are also associated with more severe disease, low diffusing capacity for carbon monoxide (DLCO), and advanced-stage nailfold capillaroscopy changes in patients with scleroderma [67, 68]. However, a recent meta-analysis revealed that though VDD is quite common in scleroderma patients, it does not correlate to the disease activity [69].

Over the years, it has been proven that vitamin D is necessary for optimum muscle and bone health. In CTDs, vitamin D levels correlate with intensity of muscle weakness [68, 70]. It may also be considered as one of the risk factors in developing myositis [71]. However, the role of vitamin D in myositis or other CTDs has not been established yet. Studies have shown that in fibromyalgia, VDD is correlated with pain and disease activity [72] and correction of deficiency improves the symptoms [73].

11. Undifferentiated arthritis

Significant association has been reported between vitamin D deficiency and nonspecific musculoskeletal pain, arthralgias, or undifferentiated arthritis [74, 75]. A positive correlation of VDD with undifferentiated arthritis [76] and early inflammatory arthritis [77] has been observed. It has also shown VDD as one of the risk factors of disease progression to RA [76, 78].

Author details

Binit Vaidya* and Shweta Nakarmi
National Center for Rheumatic Diseases, Kathmandu, Nepal

*Address all correspondence to: drbinitvaidya@yahoo.com

References

[1] Bikle DD. Vitamin D metabolism, mechanism of action, and Clinical Applications. Chemistry and Biology. 2014;**21**(3):319-329

[2] Kennel KA, Drake MT, Hurley DL. Vitamin D deficiency in adults: When to test and how to treat. Mayo Clinic Proceedings. 2010;**85**(8):752-758

[3] Holick MF. Sunlight and vitamin D for bone health and prevention of autoimmune diseases, cancers, and cardiovascular disease. The American Journal of Clinical Nutrition. 2004;**80** (6 Suppl):1678S-1688S

[4] Nussey S, Whitehead S. The parathyroid glands and vitamin D. In: Endocrinology: An Integrated Approach. England, Oxford: BIOS Scientific Publishers; 2001

[5] Ebeling PR. Vitamin D and bone health: Epidemiologic studies. BoneKEy Reports. 2014;**3**:511. DOI: 10.1038/bonekey.2014.6 Published: 5 March 2014

[6] Holick MF, Binkley NC, Bischoff-Ferrari HA, et al. Evaluation, treatment, and prevention of Vitamin D deficiency: An Endocrine Society clinical practice guideline. The Journal of Clinical Endocrinology and Metabolism. 2011; **96**(7):1911-1930

[7] Holick MF. Vitamin D: A millenium perspective. Journal of Cellular Biochemistry. 2003;**88**(2):296-307

[8] Mozolowski W. Jędrzej Sniadecki (1768-1838) on the cure of rickets. Nature. 1939;**143**(3612):121-124

[9] IOM (Institute of Medicine). Dietary Reference Intakes for Calcium and Vitamin D. Washington, DC: The National Academies Press; 2011

[10] Christakos S, Ajibade DV, Dhawan P, et al. Vitamin D: Metabolism. Endocrinology and Metabolism Clinics of North America. 2010;**39**(2):243-253

[11] Bikle DD. Vitamin D and the skin: Physiology and pathophysiology. Reviews in Endocrine & Metabolic Disorders. 2012;**13**(1):3-19

[12] Holick MF, MacLaughlin JA, Clark MB, et al. Photosynthesis of previtamin D3 in human skin and the physiologic consequences. Science. 1980;**210**(4466):203-205

[13] The British Dietetic Association. The Food Factsheet-Vitamin D. 2016. Available from: https://www.bda.uk.com/foodfacts/VitaminD.pdf

[14] Holick MF. Vitamin D deficiency. The New England Journal of Medicine. 2007;**357**:266-281

[15] Ringe JD. The effect of Vitamin D on falls and fractures. Scandinavian Journal of Clinical and Laboratory Investigation. 2012;**72**(Suppl. 243):73-78

[16] Bordelon P, Ghetu MV, Langan R. Recognition and management of Vitamin D deficiency. 2009;**80**(8): 841-846. Available from: www.aafp.org/afp

[17] Glerup H, Mikkelsen K, Poulsen L, et al. Hypovitaminosis D myopathy without biochemical signs of osteomalacic bone involvement. Calcified Tissue International. 2000; **66**(6):419-424

[18] Plotnikoff GA, Quingley JM. Prevalence of severe hypovitaminosis D in patients with persistent, nonspecific musculoskeletal pain. Mayo Clinic Proceedings. 2003;**78**(12):1463-1470

[19] Noborio R, Kobayashi K, Shintani Y, et al. Comparison of the efficacy of calcipotriol and maxacalcitol in

combination with narrow-band ultraviolet B therapy for the treatment of psoriasis vulgaris. Photodermatology, Photoimmunology & Photomedicine. 2006;**22**(5):262-264

[20] Bikle DD. Vitamin D and skin cancer. The Journal of Nutrition. 2004; **134**(12 Suppl):3472S-3478S

[21] Otani T, Iwasaki M, Sasazuki S, et al. Plasma vitamin D and risk of colorectal cancer: The Japan Public Health Center-based prospective study. British Journal of Cancer. 2007;**97**(3):446-451

[22] Jenab M, Bueno-de-Mesquita H, Ferrari P, et al. Association between pre-diagnostic circulating vitamin D concentration and risk of colorectal cancer in European populations: A nested case-control study. BMJ. 2010: 340, b5500

[23] Van Etten E, Mathieu C. Immunoregulation by 1,25-dihydroxyvitamin D3: Basic concepts. The Journal of Steroid Biochemistry and Molecular Biology. 2005;**97**(1–2):93-101

[24] Daniel C, Sartory NA, Zahn N, et al. Immune modulatory treatment of trinitrobenzene sulfonic acid colitis with calcitriol is associated with a change of a T helper (Th) 1/Th17 to a Th2 and regulatory T cell profile. The Journal of Pharmacology and Experimental Therapeutics. 2008;**324**(1):23-33

[25] Munger KL, Levin LI, Hollis BW, et al. Serum 25-hydroxyvitamin D levels and risk of multiple sclerosis. JAMA. 2006;**296**(23):2832-2838

[26] Arshi S, Fallahpour M, Nabavi M, et al. The effects of Vitamin D supplementation on airway functions in mild to moderate persistent asthma. Annals of Allergy, Asthma & Immunology. 2014;**113**(4):404-409

[27] Brøndum-Jacobsen P, Benn M, Jensen GB, et al. 25-Hydroxyvitamin D levels and risk of ischemic heart disease, myocardial infarction, and early death-population-based study and meta-analyses of 18 and 17 studies. Arteriosclerosis, Thrombosis, and Vascular Biology. 2012;**32**(11): 2794-2802

[28] McGreevy C, Williams D. New insights about vitamin D and cardiovascular disease. Annals of Internal Medicine. 2011;**155**:820-826

[29] Mitri J, Dawson-Hughes B, Hu FB, Pittas AG. Effects of vitamin D and calcium supplementation on pancreatic b cell function, insulin sensitivity, and glycemia in adults at high risk of diabetes: The calcium and vitamin D for diabetes mellitus (CaDDM) randomized controlled trial. The American Journal of Clinical Nutrition. 2011;**94**(2):486-494

[30] Eyles DW, Feron F, Cui X, et al. Developmental vitamin D deficiency causes abnormal brain development. Psychoneuroendocrinology. 2009;**34** (Suppl. 1):S247-S257

[31] Adorini L, Penna G. Control of autoimmune diseases by the vitamin D endocrine system. Nature Clinical Practice Rheumatology. 2008;**4**:404-412

[32] Gatenby P, Lucas R, Swaminathan A. Vitamin D deficiency and risk for rheumatic diseases: an update. Current Opinion in Rheumatology. 2013, 2013;**25**:184-191. DOI: 10.1097/BOR.0b013e32835cfc16

[33] Adorini L, Giarratana N, Penna G. Pharmacological induction of tolerogenic dendritic cells and regulatory T cells. Seminars in Immunology. 2004;**16**(2):127-134

[34] Etten EV, Mathieu C. Immunoregulation by 1,25-dihydroxyvitamin D3: Basic concepts. Journal of Steroid Biochemistry and Molecular Biology. 2005;**97**:93-101

[35] Chen S et al. Modulatory effects of 1,25-dihydroxyvitamin D3 on human B

cell differentiation. Journal of Immunology. 2007;**179**:1634-1647

[36] Grant WB. Epidemiology of disease risks in relation to vitamin D insufficiency. Progress in Biophysics and Molecular Biology. 2006;**92**:65-79

[37] Hassanalilou T, Khalili L, Ghavamzadeh S, et al. Role of vitamin D deficiency in systemic lupus erythematosus incidence and aggravation. Autoimmunity Highlights. 2018;**9**(1). DOI: 10.1007/s13317-017-0101-x

[38] Toloza SM, Cole DE, Gladman DD, et al. Vitamin D insufficiency in a large female SLE cohort. Lupus. 2010;**19**(1): 13-19. DOI: 10.1177/0961203309345775. Epub: 6 November 2009

[39] Damanhouri LH. Vitamin D deficiency in Saudi patients with systemic lupus erythematosus. Saudi Medical Journal. 2009;**30**(10):1291-1295

[40] Szodoray P, Tarr T, Bazso A, et al. The immunopathological role of vitamin D in patients with SLE: Data from a single centre registry in Hungary. Scandinavian Journal of Rheumatology. 2011;**40**(2):122-126

[41] Bogaczewicz J, Sysa-Jedrzejowska A, Arkuszewska C, et al. Prevalence of autoantibodies directed against 1,25 (OH)2D3 inpatients with systemic lupus Erythematosus. Pol MerkurLekarski. 2010;**28**(164):103-107

[42] Mok CC, Birmingham DJ, Ho LY, et al. Vitamin D deficiency as marker for disease activity and damage in systemic lupus erythematosus: A comparison with anti-dsDNA and anti-C1q. Lupus. 2012;**21**(1):36-42

[43] Thudi A, Yin S, Wandstrat AE, et al. Vitamin D levels and disease status in Texas patients with systemic lupus erythematosus. The American Journal of the Medical Sciences. 2008;**335**(2): 99-104

[44] Patschan D, Loddenkemper K, Buttgereit F. Molecular mechanisms of glucocorticoid-induced osteoporosis. Bone. 2001;**29**:498-505

[45] Kamen DL. Vitamin D in lupus: New kid on the block? Bulletin of the NYU Hospital for Joint Diseases. 2010; **68**(3):218-222

[46] Barré PE, Gascon-Barré M, Meakins JL, Goltzman D. Hydroxychloroquine treatment of hypercalcemia in a patient with sarcoidosis undergoing hemodialysis. The American Journal of Medicine. 1987;**82**:1259-1262

[47] Sakthiswary R, Raymond AA. The clinical significance of Vitamin D in systemic lupus erythematosus: A systematic review. PLoS One. 2013;**8**(1): e55275. DOI: 10.1371/journal. pone.0055275

[48] Suzan M A, Aisha M S. Vitamin D deficiency in patients with systemic lupus erythematosus. Oman Medical Journal. 2013;**28**(1):42-47. DOI: 10. 5001/omj.2013.10

[49] Amital H, Szekanecz Z, Szucs G, et al. Serum concentrations of 25-OH vitamin D in patients with systemic lupus erythematosus (SLE) are inversely related to disease activity: Is it time to routinely supplement patients with SLE with vitamin D? Annals of the Rheumatic Diseases. 2010;**69**(6): 1155-1157

[50] Yap KS, Northcott M, AB-Y H, et al. Association of low vitamin D with high disease activity in an Australian systemic lupus erythematosus cohort. Lupus Science and Medicine. 2015;**2**: e000064. DOI: 10.1136/lupus2014-000064

[51] Ruiz-Irastorza G, Gordo S, Olivares N, et al. Changes in vitamin D levels in patients with systemic lupus erythematosus: Effects on fatigue, disease activity, and damage. Arthritis

Care & Research (Hoboken). 2010; **62**(8):1160-1165

[52] Przybelski RJ, Binkley NC. Is vitamin D important for preserving cognition? A positive correlation of serum 25-hydroxyvitamin D concentration with cognitive function. Archives of Biochemistry and Biophysics. 2007;**460**(2):202-205

[53] Deane K, Demoruelle MK, Kelmenson LB, et al. Genetic and environmental risk factors for rheumatoid arthritis. Best Practice & Research. Clinical Rheumatology. 2017; **31**(2017):3-18

[54] Tetlow LC, Woolley DE. 1999 The effects of 1α,25-dihydroxyvitamin D3 on matrix metalloproteinase and prostaglandin E2 production by cells of the rheumatoid lesion [peer-reviewed primary research]. Available from: http://arthritisresearch.com/14oct99/ar0101p01

[55] Tsirogianni A, Hadjicostas P. AB0821 high prevalence of vitamin d3 deficiency in patients with rheumatic diseases and musculoskeletal disorders in cyprus. Annals of the Rheumatic Diseases. 2017;**76**:1345

[56] Cecchetti S, Tatar Z, Galan P, et al. Prevalence of vitamin D deficiency in rheumatoid arthritis and association with disease activity and cardiovascular risk factors: data from the COMEDRA study. Clinical and Experimental Rheumatology. 2016;**34**(6):984-990. Epub: 30 September 2016

[57] Meena N, Chawla SPS, Garg R, et al. Assessment of vitamin D in rheumatoid arthritis and its correlation with disease activity. Journal of Natural Science, Biology and Medicine. 2018, 2018;**9**(1): 54-58. DOI: 10.4103/jnsbm.JNSBM_128_17

[58] Lee YH, Bae SC. Vitamin D level in rheumatoid arthritis and its correlation with the disease activity: A meta-analysis. Clinical and Experimental Rheumatology. 2016;**34**:827-833

[59] Lin J, Liu J, Davies ML, Chen W. Serum vitamin D level and rheumatoid arthritis disease activity: Review and meta-analysis. PLoS One. 2016;**11**(1): e0146351. DOI: 10.1371/journal.pone.0146351

[60] Welsh P, Peters MJ, McInnes IB, et al. Vitamin D deficiency is common in patients with RA and linked to disease activity, but circulating levels are unaffected by TNFα blockade: Results from a prospective cohort study. Annals of the Rheumatic Diseases. 2011;**70**(6): 1165-1167. DOI: 10.1136/ard.2010.137265. Epub: 3 November 2010

[61] Urruticoechea-Arana et al. Vitamin D deficiency in chronic inflammatory rheumatic diseases: Results of the cardiovascular in rheumatology [CARMA] study. Arthritis Research and Therapy. 2015;**17**:211. DOI: 10.1186/s13075-015-0704-4

[62] Kostoglou-Athanassiou I, Athanassiou P, Lyraki A, et al. Vitamin D and rheumatoid arthritis. The rAdvEndocrinolMetab. 2012;**3**(6): 181-187. DOI: 10.1177/2042018812471070

[63] Feser M, Derber LA, Deane KD, et al. Plasma 25, OH vitamin D levels are not associated with rheumatoid arthritis-related autoantibodies in individuals at elevated risk for rheumatoid arthritis. The Journal of Rheumatology. 2009;**36**(5):943-946. DOI: 10.3899/jrheum.080764.

[64] Hajjaj-Hassouni N, Mawani N, Allali F, et al. Evaluation of vitamin D status in rheumatoid arthritis and its association with disease activity across 15 countries: "The COMORA Study". International Journal of Rheumatology. 2017;**2017**:8. DOI: 10.1155/2017/5491676. Article ID 5491676

[65] Zold E, Szodoray P, Gaal J, et al. Vitamin D deficiency in undifferentiated connective tissue disease. Arthritis Research & Therapy. 2008;**10**:R123

[66] Artaza JN, Norris KC. Vitamin D reduces the expression of collagen and key profibrotic factors by inducing an antifibrotic phenotype in mesenchymal multipotent cells. The Journal of Endocrinology. 2009;**200**:207-221

[67] Zold E, Barta Z, Bodolay E. Vitamin D deficiency and connective tissue disease. Vitamins and Hormones. 2011: 261-286. DOI: 10.1016/b978-0-12-386960-9.00011-3

[68] Groseanu L, Bojinca V, Gudu T, et al. Low vitamin D status in systemic sclerosis and the impact on disease phenotype. European Journal of Rheumatology. 2016;**3**(2):50-55. DOI: 10.5152/eurjrheum.2015.0065

[69] An L, Sun MH, Chen F, Li JR. Vitamin D levels in systemic sclerosis patients: A meta-analysis. Drug Design, Development and Therapy. 2017;**11**: 3119-3125. DOI: 10.2147/DDDT. S144860. Published: 27 October 2017

[70] Tanner SB, Harwell SA. More than healthy bones: A review of vitamin D in muscle health. Advances in Musculoskeletal Disease. 2015;**7**(4): 152-159. DOI: 10.1177/1759720X 15588521

[71] Azali P, Barbasso Helmers S, Kockum I, et al. Low serum levels of vitamin D in idiopathic inflammatory myopathies. Annals of the Rheumatic Diseases. 2013;**72**(4):512-516. DOI: 10.1136/annrheumdis-2012-201849 Epub: 19 Septemper 2012

[72] Makrani AH, Afshari M, Ghajar M, et al. Vitamin D and fibromyalgia: A meta-analysis. Korean Journal of Pain. 2017;**30**(4):250-257. DOI: 10.3344/ kjp.2017.30.4.250

[73] de Carvalho JF, da Rocha Araújo FAG, da Mota LMA, et al. Vitamin D supplementation seems to improve fibromyalgia symptoms: Preliminary results. The Israel Medical Association Journal. 2018;**20**(6):379-381

[74] Heidari B, Heidari P, Tilaki KH. Relationship between unexplained arthralgia and vitamin D deficiency: A case control study. Acta Medica Iranica. 2014;**52**(5):400-405

[75] Sheikh NI, Farooq R. Arthralgia an indicator of vitamin D3 deficiency and insufficiency. JIMDC. 2017;**6**(4)

[76] Heidari B, Hajian-Tilaki K, Heidari P. The status of serum vitamin D in patients with rheumatoid arthritis and undifferentiated inflammatory arthritis compared with controls. Rheumatology International. 2012; **32**(4):991-995. DOI: 10.1007/ s00296-010-1736-3. Epub: 19 January 2011

[77] Park YE, Kim BH, Lee SG, et al. Vitamin D status of patients with early inflammatory arthritis. Clinical Rheumatology. 2015;**34**(2):239-246

[78] Lee SG, Kim GT, Lee JW, et al. Vitamin D deficiency in patients with early inflammatory arthritis. In: 2013 ACR/ARHP Annual Meeting; Abstract number 394. 2013

Chapter 7

Extraskeletal Effects of Vitamin D Deficiency in Intensive Care Patients

Edward T. Zawada

Abstract

There has been a progression of investigations of the biology and pharmacology of vitamin D. Original work proved the importance of this axis on bone and skeletal homeostasis. Subsequent knowledge of basic cellular physiology led to studies of the role of vitamin D in other tissues where calcium flux is important to cellular functions. This was the beginning of exploration of vitamin D in extraskeletal health. Next came investigations into extraskeletal diseases which appeared to be more prevalent in vitamin D-deficient individuals. It was surprising that not only was there higher mortality from those diseases associated with low vitamin D levels, but also all-cause mortality was higher as well. Cellular pathophysiology of these findings was explored. Finally, studies have focused on outcomes in seriously ill patients with those diseases when hospitalized in the intensive care unit (ICU). Inverse correlations have been seen of several common ICU outcomes with levels of vitamin D at entry to the ICU, but the current effort is now in clarifying a role for routine measurement of these levels and the possible role of at least vitamin D replacement or even supplementation in the ICU patient with multiple organ pharmacologic or mechanical life support.

Keywords: vitamin D, extraskeletal, supplementation

1. Introduction

The important role of vitamin D in calcium homeostasis which in turn is so important to the cellular functions of diverse organs naturally led to interest in studying its role in extraskeletal health [1]. Calcium ions play an important role in signal transduction pathways where they act as a second messenger, in neurotransmitter release from neurons, in contraction of all muscle cell types, and in fertilization. It was natural for studies to begin to explore the impact of vitamin D on vascular tone, cardiac contractility, neuron function, and hormone release and function. As mentioned in the introductory chapter, my observations led to further studies on the role of vitamin D and calcium on vascular tone, blood pressure, and hypertensive states [2, 3]. It was not surprising when Framingham data and the National Health and Nutrition Examination Survey (NHANES 2001–2004) suggested that individuals with low levels of vitamin D were more likely to have a cardiovascular event, a higher prevalence of angina and myocardial infarction, and higher prevalence of heart failure and peripheral arterial disease [4, 5]. Reports of the consequences of vitamin D on mortality in a variety of serious illnesses and even

IntechOpen

all-cause mortality were furthering this expanded role for vitamin D [5]. The purpose of this chapter is to review the evidence for and against the impact of deficiency of vitamin D on the most seriously ill patients, those in the intensive care unit (ICU).

2. Extraskeletal dysfunction in vitamin D deficiency

A brief menu of organ dysfunction previously reported is now presented as a prelude to understanding possible contribution to prolonged recovery or higher mortality in seriously ill ICU patients [1].

a. Muscle weakness: some descriptive studies have associated muscle weakness with vitamin D deficiency. For example, vitamin D-deficient children have severe muscle weakness. However, randomized studies of supplementation have only led to some improvement in these series.

b. Falls due to musculoskeletal dysfunction has been reported to be associated with increased vitamin D-deficient elders with benefit of supplementation seen in many studies, especially in those patients with the lowest 25-hydroxy vitamin D levels, < 25 nmol/L.

c. Cancer incidence is associated with vitamin D deficiency for patients with colon cancer and breast cancer in postmenopausal women, but not prostate cancer. There are many small series of worsening outcomes in hematologic malignancies, especially in children, in vitamin D-deficient individuals, but supplementation has not been clearly shown to improve outcomes. Since many of these patients are admitted to the ICU for short-term management of emergency complications of their diseases or treatments, surveillance of vitamin D may be important to the management of these patients. A very recent report on the subject of vitamin D supplementation and colorectal cancer will be presented in the Conclusion section of this chapter.

d. Immune system: vitamin D has effects on all cells of the immune system, and there are reports of deficiency associated with such immune-mediated diseases as diabetes, multiple sclerosis, and inflammatory bowel disease. Certainly in patients with these medical problems, complications develop requiring admission to the ICU. So it would seem that more studies and larger numbers of patients in these series are needed in the monitoring and at least maintenance of normal levels in patients with these common immune-mediated diseases.

e. Asthma: severe bronchospasm is a common reason for admission to the ICU, often in a younger or pediatric patient. The data on the role of vitamin D in this problem thus far is considered inconclusive. There has also been no good evidence of the need to aggressively search or supplement vitamin D in patients with chronic obstructive pulmonary disease (COPD).

f. Infection: despite the evidence for vitamin D on the immune system, no specific infections have required correction or supplementation with Vitamin D to improve success in the management of seriously ill patients.

g. CV system: in my introduction to this book, I review my previous work on the role of calcium and vitamin D in vascular tone and blood pressure [2, 3]. Despite these reports, no conclusive evidence has led to routine screening, correction, or supplementation of vitamin D to prevent or control hypertension. In the Framingham Study [4], a relationship of vitamin D deficiency and risk of cardiovascular events was raised. This concern was furthered by data from the National Health and Nutrition Education Survey (NHANES) data [5].

h. Diabetes: as alluded previously there have been studies of increased incidence and severity of type 1 diabetes mellitus (T1DM) in vitamin D populations, but the evidence has not been conclusive. Diabetes is likely the most common comorbidity in intensive care unit patients.

i. Neuropsychiatry: despite the relationship of calcium and vitamin D to membrane transport and neurohumoral secretion, the impact of vitamin D or its supplementation or augmentation on ICU neurologic diseases such as stroke or seizures has not been proven.

j. Pregnancy: there are reports of more complications with vitamin D-deficient mothers. No definitive studies suggest a role for monitoring or replacement or augmentation in ICU syndromes in pregnancy such as preeclampsia or hemolysis, elevated liver enzymes, and low platelets (HELLP) syndrome.

3. Studies of vitamin D specific to patients in the ICU

There are multiple publications concerning vitamin D deficiency affecting mortality in seriously ill patients [6, 7]. Whereas these studies are prospective and controlled, the numbers of patients have not been great. In a report by Han et al. [7] high-dose vitamin D administration was studied in ventilated intensive care unit patients. This was a small pilot double-blinded randomized controlled trial. Twelve subjects were in each group. The levels were low in 43% of patients. The groups were given 50,000 vs. 1,000,000 units daily enterally vs. placebo for 5 days. High-dose vitamin D3 increased levels of 25OHD and led to decreased numbers of hospital days without any other improved clinical outcomes. The decrease in hospital length of stay not ICU days or ventilator days or hospital-acquired infections, hospital mortality, or mortality at day 84 was seen.

There are additional reports of the role of vitamin D in patients with specific major organ dysfunction. These reports have studied heart, lung, or kidney failure in the ICU. Those studies which have dealt with vitamin D supplements to assist with patients to be liberated from the ventilator because of respiratory failure have shown no benefit of supplementation to reduce time on the ventilator and speed liberation [7].

There are a considerable number of journal articles dealing with cardiac function in patients with disturbed calcium metabolism [8] and improvement of heart failure with vitamin D administration [9]. Renal failure is another common organ failure developing in seriously ill patient who requires often intensive care. There is a long tradition of vitamin D administration to dialysis patients due to their impaired renal activation of 25OH vitamin D to 1,25 dihydroxyvitamin D. New findings from a real-world study of 52,757 patients confirmed that vitamin D improves survival in hemodialysis patients [10].

4. Contradictory recent major studies

In the past few decades, there appeared to be a crescendo of possible multisystem benefits from vitamin D supplementation reported throughout the world medical literature. However, the crescendo has been halted by several recent negative studies. In a long-awaiting randomized, double-blinded, controlled trial, Manson studied vitamin D dosing of 2000 units of D3 daily in 25,000 adults with a mean age of 67 followed for a mean of 5.3 years [11]. He did not show difference in the occurrence of cardiovascular diseases or outcome of patients with invasive cancer or incidence of death from cancer, cardiovascular disease, or any cause which did not differ from those treated with placebo.

However, another recent prospective placebo-controlled report by Witte et al. [12] still suggests usefulness of vitamin D supplementation in patients with heart failure which likely contributes to morbidity in many if not most ICU patients. This is the vitamin D treating patients with chron***ic*** he***art*** failur**e** trial (VINDICATE). In this trial, there was no improvement in a 6-min walk but had beneficial effects on left ventricular structure and function in 229 patients with vitamin D levels <50 nmol/L or which is <20 ng/mL treated with 4000 units daily who had already been on contemporary optimal medical therapy for their heart failure with optimal medical therapy. Left ventricular structure was assessed by changes in stroke volume, and remodeling was assessed from improved left ventricular systolic diameter and end-diastolic diameter. One of the most recent issues of JAMA reported that high-dose vitamin D vs. standard-dose vitamin D when added to standard chemotherapy had no benefit for the outcome of median progression-free survival but did reduce the hazard ratio for progression or death. [13].

Just when the pendulum in the world's literature seems to swing in one direction, new publications are reported for the opposite conclusions about vitamin D and extraskeletal health. The pendulum right now has swung against routine screening or supplementation in the treatment or prevention of extraskeletal organ dysfunction. But then Witte and the recent studies in dialysis patients suggest that the patients with heart and kidney dysfunction do benefit not only in organ function but also overall survival.

5. Conclusions and recommendations

There is some evidence that deficiency of vitamin D at the time of entry to the ICU is associated with worse outcomes. There are weak correlations between vitamin D levels and outcomes during the course of the ICU stay. Vitamin D is not recommended at this time to be routinely checked in ICU patients unless part of several large studies is ongoing worldwide such as the VINDICATE trial described above. Supplementation with vitamin D is routine in patients with renal failure and does improve outcomes in that group of patients commonly hospitalized in the ICU. There appears some benefit to supplementation of vitamin D in patients with heart failure, another group commonly hospitalized in the ICU. Studies are ongoing in this subgroup of patients. There appears no evidence for measurement or supplementation of vitamin D to patients with COPD exacerbations, asthma, and pneumonia or to assist with liberation from the ventilator. More large-scaled, multicenter, prospective, randomized, and controlled trials are needed. Dosing duration and correction of deficiency vs. creation of elevated levels by low- or high-dose treatments are all variables which need systematic study.

Author details

Edward T. Zawada
Department of Internal Medicine, University of South Dakota, Sanford School of Medicine, USA

*Address all correspondence to: ezawada@sio.midco.net

References

[1] Bouillon R. Vitamin D and extraskeletal health. UpToDate. 2019:1-22

[2] Zawada ET Jr, TerWee J, McClung DE. Systemic and renal vascular responses to dietary calcium and vitamin D. Hypertension. 1988;**8**:975-982

[3] Zawada ET Jr, editor. Introduction—Vitamin D Facts and Fads. London, England: InTech Publishers. 2019

[4] Wang TJ, Pencina MJ, Booth SL, et al. Vitamin D deficiency and risk of cardiovascular disease. Circulation. 2008;**117**:503-511

[5] Kendrik J, Targher G, Smits G, Chonchol M. 2-hydroxyvitamin D deficiency is independently associated with cardiovascular disease in the Third National Health and Nutrition Examination Survey. Atherosclerosis. 2009;**205**:255-260

[6] Moraes RB, Friedman G, Wawrzeniak IC, Marques LS, Nagel FM, Lisboa TC, et al. Vitamin D deficiency is independently associated with mortality among critically ill patients. Clinics (São Paulo, Brazil). 2015;**70**:326-332

[7] Han JE, Jones JL, Tangoricha V, Brown MA, Hao L, Hebbar G, et al. High dose vitamin D administration in ventilated intensive care unit patients: A pilot double blind randomized controlled trial. Journal of Clinical & Translational Endocrinology. 2016;**4**:59-65

[8] Jensen A-SC, Polcwiartek C, Segaard P, Mortensen RN, Davidsen L, Aldahl ME, et al. The association between serum calcium and short-term mortality in patients with chronic heart failure. American Journal of Medicine. 2019;**132**(2):200-208

[9] Trivedi DP, Doll R, Khaw KT. Effect of four monthly oral vitamin D3(cholecalciferol) supplementation on fractures and mortality in men and women living in the community: Randomized double blind controlled trial. British Medical Journal. 2003;**326**(7387):469

[10] Kao YY-H et al. Active vitamin D therapy prolongs survival in hemodialysis. In: Persaud N, editor. Renal and Urology News. 2019. p. 1

[11] Manson JE, Cook NR, Lee IM, et al. Vitamin D supplements and prevention of cancer and cardiovascular disease. The New England Journal of Medicine. 2019;**30**:33

[12] Witte KK, Byron R, Gierula J, Paton ME, Jami HA, Lowry JE, et al. Effects of vitamin D on cardiac function in patients with chronic heart failure: The VINDICATE study. Journal of the American College of Cardiology. 2016;**67**(22):2593-2603

[13] Ng K et al. Effect of high-dose vs. standard dose vitamin D3 supplementation on progression-free survival among patients with advanced or metastatic colorectal cancer. The SUNSHINE randomized clinical trial. JAMA. 2019;**321**(14):1370-1379